Human-in-the-Loop

Human-in-the-Loop
Probabilistic Modeling of an Aerospace Mission Outcome

Ephraim Suhir

CRC Press is an imprint of the
Taylor & Francis Group, an **informa** business

CRC Press
Taylor & Francis Group
6000 Broken Sound Parkway NW, Suite 300
Boca Raton, FL 33487-2742

© 2018 by Taylor & Francis Group, LLC
CRC Press is an imprint of Taylor & Francis Group, an Informa business

No claim to original U.S. Government works

Printed on acid-free paper

International Standard Book Number-13: 978-0-8153-5455-0 (Hardback)

This book contains information obtained from authentic and highly regarded sources. Reasonable efforts have been made to publish reliable data and information, but the author and publisher cannot assume responsibility for the validity of all materials or the consequences of their use. The authors and publishers have attempted to trace the copyright holders of all material reproduced in this publication and apologize to copyright holders if permission to publish in this form has not been obtained. If any copyright material has not been acknowledged please write and let us know so we may rectify in any future reprint.

Except as permitted under U.S. Copyright Law, no part of this book may be reprinted, reproduced, transmitted, or utilized in any form by any electronic, mechanical, or other means, now known or hereafter invented, including photocopying, microfilming, and recording, or in any information storage or retrieval system, without written permission from the publishers.

For permission to photocopy or use material electronically from this work, please access www. copyright.com (http://www.copyright.com/) or contact the Copyright Clearance Center, Inc. (CCC), 222 Rosewood Drive, Danvers, MA 01923, 978-750-8400. CCC is a not-for-profit organization that provides licenses and registration for a variety of users. For organizations that have been granted a photocopy license by the CCC, a separate system of payment has been arranged.

Trademark Notice: Product or corporate names may be trademarks or registered trademarks, and are used only for identification and explanation without intent to infringe.

Library of Congress Cataloging-in-Publication Data

Names: Suhir, Ephraim, author.
Title: Human-in-the-loop : probabilistic modeling of an aerospace mission
outcome / Ephraim Suhir.
Description: First edition. | Boca Raton, FL : CRC Press/Taylor & Francis Group,
2018. | "A CRC title, part of the Taylor & Francis imprint, a member of the
Taylor & Francis Group, the academic division of T&F Informa plc." | Includes
bibliographical references and index.
Identifiers: LCCN 2017055664 | ISBN 9780815354550 (hardback : acid-free paper) |
ISBN 9781351132510 (ebook)
Subjects: LCSH: Aeronautics--Human factors. | Systems engineering--Simulation
methods. | Probabilities.
Classification: LCC TL553.6 .S84 2018 | DDC 629.1--dc23
LC record available at https://lccn.loc.gov/2017055664

Visit the Taylor & Francis Web site at
http://www.taylorandfrancis.com

and the CRC Press Web site at
http://www.crcpress.com

Contents

Preface ix

Author xi

1. Introduction: Probabilistic Modeling Approach in Aerospace Engineering 1

2. Fundamentals of Applied Probability 9
    2.1   Random Events 9
    2.2   Algebra of Events 10
    2.3   Discrete Random Variables 22
            2.3.1   Probability Characteristics 22
            2.3.2   Poisson Distribution 26
    2.4   Continuous Random Variables 29
            2.4.1   Probability Characteristics 29
            2.4.2   Bayes Formula for Continuous Random Variables 31
            2.4.3   Bayes Formula as a Technical Diagnostics Tool 33
            2.4.4   Uniform Distribution 38
            2.4.5   Exponential Distribution 39
            2.4.6   Normal (Gaussian) Distribution 40
            2.4.7   Rayleigh Distribution 42
            2.4.8   Weibull Distribution 44
            2.4.9   Beta-Distribution 45
            2.4.10 Beta-Distribution as a Tool for Updating Reliability Information 46
    2.5   Functions of Random Variables 51
    2.6   Extreme Value Distributions 54

3. Helicopter-Landing-Ship and the Role of the Human Factor 57
    3.1   Summary 57
    3.2   Introduction 58
    3.3   Probability That the Operation Time Exceeds a Certain Level 59
    3.4   Probability That the Duration of Landing Exceeds the Duration of the Lull 63
    3.5   The Probability Distribution Function for the Extreme Vertical Velocity of Ship's Deck 64
    3.6   Allowable Landing Velocity When Landing on a Solid Ground 65
    3.7   Allowable Landing Velocity When Landing on a Ship Deck 66
    3.8   The Probability of Safe Landing on a Ship's Deck 68
    3.9   Conclusions 70
    References 70

v

Contents

4. Fundamentals of Probabilistic Aerospace Electronics Reliability Engineering 73

4.1 Today's Practices: Some Problems Envisioned and Questions Asked 73

4.2 Accelerated Testing 74

4.3 PDfR and Its Major Principles ("10 Commandments") 77

4.4 FOAT ("Transparent Box") as an Extension of HALT ("Black Box") 79

4.5 Design for Reliability of Electronics Systems: Deterministic and Probabilistic Approaches 80

4.6 Two Simple PDfR Models 81

4.7 BAZ Model: Possible Way to Quantify Reliability 84

4.8 Multiparametric BAZ Model 85

4.9 The Total Cost of Reliability Could Be Minimized: Elementary Example 87

4.10 Possible Next Generation of the Qualification Tests (QTs) 89

4.11 Physics-of-Failure BAZ Model Sandwiched between Two Statistical Models: Three-Step Concept 89

4.11.1 Incentive/Motivation 89

4.11.2 Background 90

4.11.3 TSC in Modeling Aerospace Electronics Reliability 92

4.11.3.1 Step 1: Bayes Formula as a Suitable Technical Diagnostics Tool 92

4.11.3.2 Step 2: BAZ Equation as Suitable Physics-of-Failure Tool 93

4.11.3.3 Step 3: Beta-Distribution as a Suitable Reliability Update Tool 98

4.11.3.4 Step 1: Application of Bayes Formula 99

4.11.3.5 Step 2: Application of BAZ Equation 99

4.11.3.6 Step 3: Application of Beta-Distribution 103

4.12 Conclusions 103

References 103

5. Probabilistic Assessment of an Aerospace Mission Outcome 109

5.1 Summary 109

5.2 Introduction 109

5.3 DEPDF of Human Nonfailure 111

5.4 Likelihood of the Vehicular Mission Success and Safety 114

5.5 Equipment (Instrumentation) Failure Rate 116

5.6 Human Performance Failure Rate 117

5.7 Weibull Law 119

5.8 Imperfect Human versus Imperfect Instrumentation: Some Short-Term Predictions 121

5.9 Most Likely MWL 125

5.10 Most Likely HCF 126

Contents vii

5.11 Conclusions 126
References 127

6. The "Miracle-on-the-Hudson" Event: Quantitative Aftermath 131
6.1 Summary 131
6.2 Introduction 132
6.3 Miracle-on-the-Hudson Event, and the Roles of
MWL and HCF 133
6.4 Double-Exponential Probability Distribution Function 134
6.5 HCF Needed to Satisfactorily Cope with a High MWL 137
6.6 Another Approach: Operation Time versus "Available"
Landing Time 138
6.7 Miracle on the Hudson: Incident 142
6.8 Miracle on the Hudson: Flight Segments (Events) 145
6.9 Miracle on the Hudson: Quantitative Aftermath 145
6.10 Captain Sullenberger 147
6.11 Flight Attendant's Hypothetical HCF Estimate 147
6.12 UN-Shuttle Flight: Crash 149
6.13 Swissair Flight 111: Segments (Events) and Crew Errors 153
6.14 Flight 111 Pilot's Hypothetical HCF 154
6.15 Other Reported Water Landings of Passenger Airplanes 155
6.16 Conclusions 158
References 158

7. "Two Men in a Cockpit": Likelihood of a Casualty If One
of the Pilots Gets Incapacitated 161
7.1 Summary 161
7.2 Introduction 162
7.3 Double-Exponential Probability Distribution Function 164
7.4 "Two Men in a Cockpit": "Accident" Occurs When One
of the Pilots Fails to Perform His/Her Duties; "Casualty"
Occurs If They Both Fail to Do So 166
7.5 Probability of a Casualty If One of the Pilots Becomes
Incapacitated 168
7.6 Required Extraordinary (Off-Normal) versus Ordinary
(Normal) HCF Level 170
7.7 Probabilistic Assessment of the Effect of the Time of
Mishap on the Likelihood of Accident 172
7.8 Conclusions 174
References 175

8. Probabilistic Modeling of the Concept of Anticipation
in Aviation 177
8.1 Summary 177
8.2 Introduction 178

8.3	Anticipation, Its Role and Attributes	178
8.4	Probabilistic Assessment of the Anticipation Time	181
	8.4.1 Probability That the Anticipation Time Exceeds a Certain Level	181
	8.4.2 Probability That the Duration of the Anticipation Process Exceeds the Available Time	185
8.5	Probabilistic Assessment of the Success of STA from the Predetermined LTA	187
	8.5.1 Double-Exponential Probability Distribution Function	187
	8.5.2 LTA- and STA-Related MWL	190
	8.5.3 LTA- and STA-Related HCF	191
8.6	Conclusions	191
	References	192

9. Double-Exponential Probability Distribution Function for the Human Nonfailure 195

9.1	Summary	195
9.2	Introduction	196
9.3	DEPDF with Consideration of Time, Human Error, and State of Health	202
9.4	MWL	204
9.5	HCF	205
9.6	The Introduced DEPDF Makes Physical Sense	206
9.7	Underlying Reliability Physics	207
9.8	Possible FOAT-Based Procedure to Establish the DEPDF	209
9.9	Conclusions	212
	References	213

Index 219

Preface

The author's background is in several quite different and, to an extent and at first glance, rather independent areas of engineering and applied science: applied and mathematical physics, applied mechanics, materials science and engineering, electronics and photonic packaging, reliability engineering and even naval architecture. When they wrote an article of me in Bell Labs *News*, the subtitle of this article was "an engineer moves from ships to chips." About 8 years ago, I "moved" again, this time from chips to the "human-in-the-loop" field that deals with tasks, missions, and situations, when a human's role is critical, and when an equipment-and-instrumentation (both hardware and software) performance contributes jointly to the outcome of a critical task, mission, or an extraordinary situation. My objective is to demonstrate in this book that some aerospace psychology-related problems lend themselves to quantification and, because nobody and nothing is perfect, that such a quantification should be done on the probabilistic basis. In effect, the difference between a practically failure-free and an insufficiently reliable human or equipment performance is merely in the level of its never-zero probability of failure. I hope that the contents of this book will provide plenty of food for thought and that many young aerospace psychologists will find some ideas developed and applied in this book useful and fruitful. I hope also that this book will add to the need to narrow the existing gap between the system-engineering and human-psychology communities in aerospace engineering. In addition, I hope that the methods and approaches suggested in this book could be applied well beyond the aerospace domain and even beyond vehicular domain in numerous cases of applied science and engineering, when human performance and equipment/instrumentation reliability contribute jointly to the outcome of a challenging effort.

This book contains nine chapters. They are based mostly on the following publications:

1. E. Suhir, "Helicopter-Landing-Ship: Undercarriage Strength and the Role of the Human Factor,"*ASME OMAE Journal*, vol. 132, No.1, December 22, 2009.

2. E. Suhir and R.H. Mogford, "'Two Men in a Cockpit': Probabilistic Assessment of the Likelihood of a Casualty if One of the Two Navigators Becomes Incapacitated," *Journal of Aircraft*, vol. 48, No. 4, July–August 2011.

3. E. Suhir, "Likelihood of Vehicular Mission-Success-and-Safety," *Journal of Aircraft*, vol. 49, No. 1, 2012.

4. E. Suhir, "Could Electronics Reliability Be Predicted, Quantified and Assured?" *Microelectronics Reliability*, No. 53, April 15, 2013.

5. E. Suhir, "'Miracle-on-the-Hudson': Quantified Aftermath," *Internati onal Journal of Human Factors Modeling and Simulation*, April 2013.

6. E. Suhir, "Human-in-the-Loop (HITL): Probabilistic Predictive Modeling (PPM) of an Aerospace Mission/Situation Outcome," *Aerospace*, No. 1, 2014.

7. E. Suhir, "Three-Step Concept in Modeling Reliability: Boltzmann-Arrhenius-Zhurkov Physics-of-Failure-Based Equation Sandwiched Between Two Statistical Models," *Microelectronics Reliability*, October 2014.

8. E. Suhir, "Human-in-the-Loop: Probabilistic Predictive Modeling, Its Role, Attributes, Challenges and Applications," *Theoretical Issues in Ergonomics Science (TIES)*, published online, July 2014.

9. E. Suhir, C. Bey, S. Lini, J.-M. Salotti, S. Hourlier, and B. Claverie, "Anticipation in Aeronautics: Probabilistic Assessments," *Theoretical Issues in Ergonomics Science*, published online, June 2014.

10. J.-M. Salotti and E. Suhir, "Manned Missions to Mars: Minimizing Risks of Failure,"*Acta Astronautica*, vol. 93, January 2014.

11. E. Suhir, "Analytical Modeling Occupies a Special Place in the Modeling Effort," Short Communication, *Journal of Physics and Mathematics*, vol. 7, No. 1, 2016.

12. E. Suhir, "Human-in-the-Loop: Application of the Double Exponential Probability Distribution Function Enables One to Quantify the Role of the Human Factor," *International Journal of Human Factor Modeling and Simulation*, vol. 5, No. 4, 2017.

Author

Ephraim Suhir is a Foreign Full Member (Academician) of the National Academy of Engineering, Ukraine. He is a Life Fellow of the Institute of Electrical and Electronics Engineers, the American Society of Mechanical Engineers, the International Microelectronics Assembly and Packaging Society, and the International Society for Optical Engineering; Fellow of the American Physical Society, the Institute of Physics (the United Kingdom), the Society of Plastics Engineers; and Associate Fellow of the American Institute of Aeronautics and Astronautics. He has authored more than 400 publications (patents, technical papers, book chapters, and books), taught courses, and presented numerous keynote and invited talks worldwide. He has received many professional awards, including in 2004 the American Society of Mechanical Engineers Worcester Read Warner Medal for outstanding contributions to the permanent literature of engineering, being the third "Russian American," after S. Timoshenko and I. Sikorsky, to receive this prestigious award, and in 1996 the Bell Laboratories Distinguished Member of Technical Staff Award for developing effective methods for predicting the reliability of complex structures employed in AT&T and Lucent Technologies products.

1

Introduction: Probabilistic Modeling Approach in Aerospace Engineering

Improvements in safety in the air and in outer space can be achieved through better ergonomics, better work environment, and other efforts of the traditional avionic psychology and ergonomic science that directly affect human behaviors and performance. There is, however, also a significant potential for further reduction in aerospace accidents and casualties, and, more importantly, for assuring the success and safety of an aerospace flight or a mission, through better understanding the roles that various uncertainties play in the planner's and operator's worlds of work, when never-perfect human, never completely failure-free navigation equipment and instrumentation, never 100%-predictable response of the object of control (air- or spacecraft), and uncertain and often harsh environments contribute jointly to the likelihood of a mishap in air or in space, and/or to the assurance of a failure-free flight, mission, or an extraordinary situation in the air or in space. By employing quantifiable and measurable ways of assessing the role and significance of various critical uncertainties and treating a human-in-the-loop (HITL) as a part, often the most crucial part, of a complex man-instrumentation-equipment-vehicle-environment system, one could improve dramatically the state of the art in assuring aerospace mission success and operational safety. This can be done by predicting, quantifying, and, if appropriate and possible, even specifying an adequate, low enough, but never-zero probability of a mishap.

The human and the equipment-and-instrumentation performance often contribute jointly to the outcome and safety of an aerospace mission or an off-normal situation. But nobody and nothing is perfect: the difference between highly reliable humans, objects, products, or missions and insufficiently reliable ones is "merely" in the level of the never-zero probability of their failure; therefore, application of the broad and consistent probabilistic risk analysis (PRA) and probabilistic predictive modeling (PPM) provides a natural, effective, and physically meaningful way to improve the success of aerospace missions and safety in the air and in the outer space.

When success and safety are imperative, and this is always the case in aerospace engineering, as well as in military, maritime, long-haul communications, and other critical engineering systems, in which a failure-free operation is paramount, the ability to predict and to quantify the outcome of an HITL-related mission or a situation is crucial. This is not the current practice,

1

though. Today's human psychology methods are mostly qualitative and are based on accumulating and analyzing statistics. But consistent, broad, and well-substantiated application of the PRA concept, including advanced methods of this body of knowledge, can dramatically improve the state of the art in understanding and accounting for the human role and performance in various aerospace missions and situations. This application could also account for the attributes and outcomes of the interaction of the never-perfect human with the never-perfect equipment and instrumentation, with consideration of the uncertain and often harsh environmental conditions, when planning and performing a flight or an outer space mission or when anticipating or coping with an off-normal situation.

The majority of the traditional human factor–oriented human psychology and ergonomic approaches are typically based on experimentations followed by statistical analyses. The applied probability concepts addressed in this book, on the contrary, are based on, and start with, simple, physically meaningful, and flexible predictive modeling, followed, when necessary, by highly focused and cost-effective experimentations, preferably failure-oriented-accelerated testing (FOAT), geared to the chosen governing model or models. The use of such a concept enables one to quantify in advance, at the planning stage and on the probabilistic basis, the outcome of a particular HITL-related effort, mission, or situation of interest. If the predicted outcome, in terms of the most likely probability of the expected human or equipment operational failure, is not favorable and should be revisited and changed, then an appropriate sensitivity analysis (SA) based on the already developed and available methodologies and algorithms can be effectively conducted and implemented to improve the situation. The additional effort and associated times and costs will be minimal in these cases.

The solutions obtained to the problems addressed in this book are based on analytical ("mathematical") modeling, rather than on computer simulations. Analytical modeling employs mathematical methods of analysis and occupies, therefore, a special place in the modeling effort. Such modeling is able to not only come up with simple relationships that clearly indicate "what affects what," but, more importantly, can often consider and explain the underlying human psychology and physics of situations and phenomena better than computer simulation, or even experimentation, can. Computer-aided simulation has become the major tool for theoretical evaluations in ergonomics. This should be attributed to the availability of powerful and flexible computer programs, which enable one to obtain, within a reasonable time, a solution to almost any theoretical ergonomics problem. This can also be partially attributed to the illusion that computer-aided simulation is the ultimate and indispensable tool for solving any HITL-related problem. The truth of the matter is that broad application of computers has by no means made analytical approaches and solutions unnecessary or even less important. Simple and physically meaningful analytical relationships have invaluable advantages because of the clarity and compactness of the information

Introduction 3

and clear indication of the role of the major factors affecting the given ergonomic phenomenon or navigational device and computer performance. But even when no difficulties are encountered in the application of simulation programs and techniques in ergonomic science problems, it is always advisable to investigate the problem analytically in addition to carrying out a simulation. Such a preliminary investigation helps to reduce computer time and expense, develop the most feasible and effective simulation model, and, in many cases, avoid fundamental errors. Clearly, if the simulation data are in agreement with the results of an analytical modeling (moreover that such modeling uses a prior, rather than a posterior approach, and might be based on different assumptions), then there is a reason to believe that the obtained solution is accurate enough.

One important feature of analytical modeling is that the same formalism developed for a particular applied science problem might be applicable to many other, even physically completely different problems. A classic example is the application of the Boltzmann's formula in the kinetic theory of gases to the, also classic, Arrhenius theory of chemical reactions. Then, there is the application of Arrhenius' theory to the reliability physics and fracture mechanics of materials (Zhurkov's model), including semiconductor materials. And recently, the application of the Boltzmann–Arrhenius–Zhurkov (BAZ) multiparametric model in various reliability physics problems of materials and systems subjected to various, not necessarily mechanical, stresses is suggested. But even within the framework of the book, the convolution of the normal and double-exponential distribution (extreme value–type distribution based on the Rayleigh distribution) was applied first by the author of this book to the under-keel clearance problem for supertankers passing through shallow waterways; then, with some modifications, to the helicopter-landing-ship (HLS) situation with consideration of the role of the human factor (HF); and then, with additional modifications, to the concept of anticipation in aerospace psychology.

A crucial requirement for an effective analytical model is its simplicity. A good analytical model should be based on physically meaningful considerations and produce simple relationships, clearly indicating the role of the major factors affecting the phenomenon or the object or the technology of interest. One authority in applied physics remarked, perhaps only partly in jest, that the degree of understanding of a phenomenon or a situation is inversely proportional to the number of variables used for its description. We deliberately used in different chapters of this book similar formulas and even the same notation when addressing quite different HITL-related situations. In our judgment, it will relieve to some extent the reader's effort, which, considering the scope of this book, will not be easy at all, to say the least.

Another important feature of a simple, easy-to-use, physically meaningful, and hence, highly practical predictive model, whether deterministic or probabilistic, is that as long as such a model is established and agreed upon, the statistics should be accumulated in application to and sticking with this model, and not the other way around. A good example from another field of engineering is the highly simplified model suggested about 150 years ago for calculating the overall strength of an oceangoing ship and treating her hull as a nonprismatic single beam that could statically sag or hog on a wave with length equal to the ship's length. The model is still considered, with numerous adjustments and updates, of course, by the classification societies that gear the accumulated statistics and knowledge about sea waves, inertia, and hydrodynamic forces, and so on, to this model by adjusting the effective wave height to the particular vessel and available and pertinent information. Many insightful predictive analytical models were suggested in the field of electronic materials: Boltzmann–Arrhenius' equation, used when there is evidence or belief that elevated temperature is the major cause of failure; BAZ's equation, used when the external stress and elevated temperature contribute jointly to the finite lifetime of a material or a device (and extended in this book); Eyring's equation, in which the stress is considered directly, not through the effective activation energy; Peck's equation, in which the "stress" is the relative humidity; inverse power law, such as Coffin–Manson's type of equations, which are widely used today to evaluate the low cycle fatigue lifetime of solder material; crack growth (Griffith's theory–based) equations, used to assess the fracture toughness of brittle materials (actually, the BAZ equation was also suggested by Zhurkov about 70 years ago as a fracture mechanics model); Miner–Palmgren's rule, used to quantify elastic fatigue lifetime of materials; numerous creep rate equations for materials prone to plastic deformation and creep and used when creep is important (e.g., in photonics); weakest link model, used to evaluate the time-to-failure in brittle materials with highly concentrated defects; stress–strength interference model, which is the most flexible probabilistic-design-for-reliability model that is widely used in various areas of engineering; extreme value distribution models, when there is evidence or belief that it is only the extreme stress values that essentially contribute to the finite lifetime of a material; and others.

It should be emphasized, however, that while an experimental approach, unsupported by theory, is blind, theory, not supported by an experiment, is dead. An experiment forms a basis and provides input data for a theoretical ergonomic model and then determines its viability, accuracy, and limits of application. The experiment, including specially designed and conducted experimental modeling, should be viewed as the "supreme and ultimate judge" of a theoretical model, whether simulation based or analytical. An experiment can and should often be rationally included in a theoretical and, in particular, an analytical solution. With the appropriate modifications and generalizations, the cost-effective and insightful analytical approach of this book is applicable, in the author's judgment, to numerous, not even necessarily in the aerospace and vehicular domains, HITL-related missions and situations. It could shed light on numerous tasks and phenomena when a human encounters an uncertain environment or a hazardous off-normal situation

Introduction   5

and/or is prone to make an error. The probabilistic analytical approach and its various formalisms are applicable also when there is an incentive to quantify a human's qualifications and performance, and/or when there is a need to assess and possibly improve his/her role in a particular mission or situation. It is shown in this book that the formalism initially developed in connection with the HLS task is also applicable, to a great extent, to the famous "Miracle-on-the-Hudson" event, as well as to the anticipation problem, in which short-term and long-term anticipations are considered. Here are the major principles ("10 commandments") of the approach:

1. Human capacity factor (HCF) is, in addition to the mental (cognitive) workload (MWL), an appropriate measure (not necessarily, but preferably, quantitative and not necessarily, but preferably, probability based) of the human ability to cope with an elevated MWL; the HCF could be complemented sometimes with the properly defined "human error" factor (say, by assessing, on the probabilistic basis, the mean time to failure, i.e., to error) that could be viewed as an important short-term aspect of the HCF, while the MWL could be complemented by the human state-of-health (SH) characteristic, which is also a short-term factor that could affect the level of the MWL: indeed, the highly subjective MWL can be perceived and handled differently by the human depending on his/her SH.

2. The relative levels of the MWL and HCF (whether deterministic or random) determine the probability of the human nonfailure in a particular HITL situation: if the level of the HCF is above the level of the MWL, no human failure is likely, to the same extent, as material or structural failure is likely if its strength (capacity) is above the applied stress (demand).

3. Such a probability cannot be low, but it need not be higher than necessary: it has to be adequate for a particular anticipated application, mission, or situation, and it has to be assessed in advance, and, if appropriate and possible, even specified, to make sure that the accepted probability-of-failure level is appropriate and achievable for a particular application and performers.

4. When adequate human performance is important, the ability to quantify it is imperative, especially if one intends to optimize and assure adequate human performance, with consideration of costs, time, and so on. No optimization is possible, of course, if the properties of interest are not quantified. Such a quantification could be preferably done on the probabilistic basis.

5. One cannot assure adequate human performance by conducting only traditional human psychology, mostly statistical, efforts and analyses; these might provide appreciable improvements for particular aerospace tasks but do not quantify human behavior and

performance (and, in addition, might be too and unnecessarily costly and time consuming). Adequate human performance cannot be assured by following only the existing general "best practices" not aimed at a particular situation or an application; the HITL events of interest are typically and certainly rare events and are often completely new endeavors, especially in astronautics, and the existing "best practices" either do not exist or may not be applicable.

6. To be effective, including cost effective, the predicted (anticipated) MWLs and HCFs should consider the most likely anticipated situations; obviously, the MWLs (including, perhaps, his/her SH) and the HCFs (including, perhaps, the propensity to the HE) should be different for different missions and possible situations, and should also be different for a jet fighter pilot, for a pilot of a commercial aircraft, for a helicopter pilot, for an astronaut, and so on, and should be assessed, specified, and assured differently.

7. The probabilistic approach (although not always a must: other quantitative approaches might also be applied and be useful), and particularly analytical probabilistic predictive modeling, is an effective means for improving the state of the art in the HITL field, especially in the tasks and efforts, when the never-perfect human and the never-perfect equipment (instrumentation), both hardware and software, contribute jointly to the outcome of a mission or a situation: nobody and nothing is perfect. In effect, the difference between a highly unreliable and extremely reliable human or the equipment is "merely" in the level of their never-zero probability of failure.

8. FOAT on a flight simulator or using other experimental means and equipment is viewed as an important constituent part of the PRA concept in various HITL situations: its aim is a better understanding of different psychological and physical factors underlying possible failures; it might be complemented by the Delphi effort (collecting opinions of experts); if one set out to determine the probability of human failure and the underlying reliability physics for the aerospace equipment, the FOAT type of experimentation is a must, is it not?

9. The highly popular prognostics and health management (PHM) and recently suggested probabilistic design for reliability (PDfR) of electronic, photonic, and microelectronic and mechanical systems (MEMS) materials and systems could and should be included in the probabilistic HITL analyses and efforts.

10. Consistent, comprehensive, and psychologically and physically meaningful probabilistic, or at least quantitative assessments, can lead to the most feasible HITL qualification (certification)

Introduction   7

methodologies, practices, and specifications, thereby considerably improving the state of the art in the field of aerospace human psychology.

These and other general HITL-related concepts are illustrated in this text by numerous examples addressing various more or less typical aerospace HITL-related problems and by providing numerous, meaningful, simple examples, mostly in Chapter 2. The objective of these examples is not only to demonstrate how the general concepts of the applied probability can be used in various HITL-related problems, but also to demonstrate how various formalisms can be employed when there is a wish or a need to quantify what is often felt intuitively, thereby raising the general "probabilistic culture" of the potential reader. These considerations explain the incentives for writing this text. It goes without saying that when the human factor is imperative, the ability to predict its role is crucial, and since nobody and nothing is perfect, this could and should be done on the probabilistic basis. The author has dealt with applied probabilities throughout his engineering career. During his tenure with the basic research area (area 11) of Bell Labs, he wrote a well-received book, *Applied Probability for Engineers and Scientists* (1997, McGraw-Hill).

2

Fundamentals of Applied Probability

2.1 Random Events

The applied theory of probability studies *random phenomena* (i.e., phenomena that do not nearly yield the same *outcomes* in repeated observations under identical conditions). Any of the possible outcomes of an *experiment (trial)* or a result of an observation is an *event*. An event is the simplest random phenomenon. Two dots appearing on the top face of a die, material or device failure, human error (such as, e.g., misinterpretation of the obtained signal) and/or his/her inability to successfully cope with an elevated mental workload (MWL), safe landing, unacceptable level of the leakage current in an electronic or optical device, and buckling of a compressed structural element, are all examples of events.

The substance of a probabilistic approach is that each event has an associated quantity that characterizes the *objective likelihood* of the occurrence of the event of interest. This quantity is called the *probability* of the event.

The classical/applied (as opposite to axiomatic/mathematical) approach to the concept of probability is as follows. If an event will inevitably occur as the result of an experiment, it is a *certain event*. An event that is known in advance as not being able to occur as a result of an experiment is an *impossible event*. The overwhelming majority of actual events are neither certain nor impossible, and may or may not occur during an experiment. These are *random events*.

Let us assume that the conditions of an experiment can be repeatedly reproduced, so that a series of identical and independent trials can be conducted, and the random event (outcome) A does or does not occur in each of these trials. If the outcome A occurred m times in n equally likely and independent trials, then the ratio $P^*(A) = \dfrac{m}{n}$ is called the *frequency*, or *statistical probability*, of the event. Obviously, $P^*(A) = 1$ for a certain event, and $P^*(A) = 0$ for an impossible event. For a random event A, $0 \leq P^*(A) \leq 1$.

When the number of experiments is small, the frequency $P^*(A)$ is an unstable quantity. It stabilizes with an increase in the number of trials and groups around a certain value $P(A)$, which is called the *probability* of the event A. If an event is expected to occur as a result of numerous repeatedly reproduced experiments, in many cases, there is no need to carry out actual experiments to determine the frequency of this event: this frequency can be predicted on the basis of the calculated probability. Such a probability is computed as the ratio of the anticipated number m of the outcomes *favorable* to the occurrence of the given event to the total number n of trials.

The content of the applied probability theory is, in effect, a collection of different ways, methods, and approaches that enable one to evaluate the probabilities of occurrence of complex events from the known or anticipated probabilities of simple ("elementary") events. This makes the probability theory different than the body of knowledge known as statistics, which is the practice of collecting and analyzing numerical data in large quantities, particularly for the purpose of inferring proportions in a whole from those in an available and supposedly representative sample. The probabilities of elementary events are often intuitively obvious or can be experimentally established using physically meaningful, highly focused, and highly cost-effective experiments. In practice, the elementary events can be often established simply by disciplines, such as, for example, applied physics, or human psychology, or ergonomics (which is the study of people's efficiency in a particular working environment). The more or less trivial examples that follow show how this could be done.

2.2 Algebra of Events

The rules of the *algebra of events* enable expression of an event of interest in terms of other events:

1. The sum $A + B$ of two events A and B is an event C consisting of the occurrence of at least one of these events. Similarly, the sum of several events $A_1, A_2, \ldots, A_n$ is the event $B = \sum_{i=1}^{n} A_i$ consisting of the occurrence of at least one of them.

2. The product AB of two events A and B is the event C consisting of simultaneous realization of both events. The product of several events $A_1, A_2, \ldots, A_n$ is the event $B = \prod_{i=1}^{n} A_i$ consisting of a simultaneous realization of all of the events.

3. It follows from the definitions of the sum and the product of events that the sums and the products of the event A with itself, with the *complete space* Ω, and with an *empty space* $\varnothing$ are

$$A + A = A, \quad A + \Omega = \Omega, \quad A + \varnothing = A, \quad AA = A, \quad A\Omega = A, \quad A\varnothing = \varnothing. \quad (2.1)$$

Fundamentals of Applied Probability 11

4. The operations of addition and multiplication of events possess the following properties:

Commutativity

$$A + B = B + A, \quad AB = BA; \tag{2.2}$$

Associativity

$$(A + B) + C = A + (B + C), \quad (AB)C = A(BC); \tag{2.3}$$

Distributivity

$$A(B + C) = AB + AC. \tag{2.4}$$

5. An *opposite* or *complementary* event of A is an event $\overline{A}$ of nonoccurrence of the event A. The domain $\overline{A}$ is the complement of the domain A with respect to the complete space Ω. As follows from the definition of the complementary event,

$$\overline{\overline{A}} = A, \overline{\Omega} = \varnothing, \overline{\varnothing} = \Omega. \tag{2.5}$$

6. The probability of the sum of two mutually exclusive events is equal to the sum of the probabilities of these events—that is, if $AB = \varnothing$, then

$$P(A + B) = P(A) + P(B). \tag{2.6}$$

This rule can be generalized for an arbitrary number of incompatible events:

$$P\left(\sum_{i=1}^{n} A_i\right) = \sum_{i=1}^{n} P(A_i). \tag{2.7}$$

As follows from this rule of the addition of probabilities, the sum of the probabilities of disjoint events $A_1, A_2, ..., A_n$ that form a complete group is equal to one. In other words, if $\sum_{i=1}^{n} A_i = \Omega$ and $A_i A_j = \varnothing$ for $i \neq j$, then

$$\sum_{i=1}^{n} P(A_i) = 1. \tag{2.8}$$

In particular, since two opposite events A and $\overline{A}$ are mutually exclusive and form a complete group, the sum of their probabilities is equal to one:

$$P(A) + P(\overline{A}) = 1. \tag{2.9}$$

Example 2.1. An outer space mission mishap is likely to happen if at least one of the three navigators makes an error. It has been established by an independent study that the probabilities of a human error during the fulfillment of the mission are 0.03, 0.04, and 0.06 for these navigators. What is the probability that no error will be made during the fulfillment of the mission, and how will this probability change, if the first navigator's human capacity factor (HCF) is so high that he/she is not expected to make an error?

The probabilities of nonfailure are 0.97, 0.96, and 0.94, and therefore the probability that no error is likely during the fulfillment of the mission is $P = 0.97 \times 0.96 \times 0.94 = 0.8753$. If the probability that the first navigator makes an error is zero, then the sought probability of nonfailure during the fulfillment of the mission will increase to $P = 1.0 \times 096 \times 0.94 = 0.9024$.

Example 2.2. The MWL G is evenly distributed between two navigators (say, between the captain and the first officer). The navigators are of the same qualifications and other HCF-related qualities. If one of the navigators gets partially or completely incapacitated, the MWL is transmitted completely to the other one. If this happens, what is the probability of nonfailure of the flight?

The flight/mission will fail, if either the captain or the first officer is unable to cope with MWL of the magnitude $\dfrac{G}{2}$ and the other navigator is unable to carry the entire MWL G. Let the probabilities of these events be $Q_{1/2}$ and Q_1, respectively. There would be no flight/mission failure in one of the following three cases:

1. None of the navigators fails when coping with the entire workload G.

2. The first mate fails under the action of the load $\dfrac{G}{2}$, but the captain is able to cope with the entire load G.

3. The captain fails under the action of the load $\dfrac{G}{2}$, but the first mate is able to cope with the entire load G.

The probabilities of nonfailures associated with these three cases are obviously $(1-Q_{1/2})^2$, $Q_{1/2}(1-Q_1)$, and $Q_{1/2}(1-Q_1)$, respectively. Then

$$P = (1-Q_{1/2})^2 + 2Q_{1/2}(1-Q_1) = 1 + Q_{1/2}^2 - 2Q_{1/2}Q_1 \tag{2.10}$$

is the probability of the flight/mission nonfailure, and the probability of its failure is

Fundamentals of Applied Probability

$$Q = 1 - P = 2Q_{1/2}Q_1 - Q_{1/2}^2. \tag{2.11}$$

Let, for example, the probability of failure of the flight/mission under the action of the elevated (extraordinarily high) MWL G be

$$Q_1 = 1 - e^{-\frac{G^2}{G_0^2}}, \tag{2.12}$$

where G_0 is the most likely (regular) MWL in an ordinary condition. If the MWL is extraordinarily high ($G \to \infty$), the probability Q_1 of failure is certainly $Q_{1,} = 1$. The probability $Q_{1/2}$ of failure under the action of half the MWL G is

$$Q_{1/2} = 1 - e^{-\frac{G^2}{4G_0^2}} = 1 - (1 - Q_1)^{1/4}, \tag{2.13}$$

and the probability of nonfailure of the flight/mission is therefore

$$P = 2(1 - Q_1)\left[1 - (1 - Q_1)^{1/4}\right] + (1 - Q_1)^{1/2}. \tag{2.14}$$

Clearly, if the HCF of both navigators is very high, the probability $Q_{1,}$ of failure to cope with the MWL of the magnitude G is zero, and the probability P of nonfailure of the flight/mission is $P = 1$. If, in the opposite situation, $Q_{1,} = 1$ (the HCF of the navigators is low), then $P = 0$. For $Q_{1,} = 0.5$ the obtained formula yields: $P = 0.1338$. This problem is revisited and addressed in detail in Chapter 7.

Example 2.3. The predicted probability that an event A will inevitably occur during the time interval T is Q. The event is equally likely to occur at any moment of time T. It has been established that the event did not occur during the time $t \leq T$. What is the probability Q_* that it will occur during the remaining time $T - t$?

The given probability Q that the event A will occur during the time interval T can be represented as the sum of the probability $\frac{t}{T}Q$ that the event A has occurred during the time t and the probability that it has not occurred during the time t. The probability of the latter event is $1 - \frac{t}{T}Q$, so that the equation

$$Q = \frac{t}{T}Q + \left(1 - \frac{t}{T}Q\right)Q, \tag{2.15}$$

should be fulfilled. Solving this equation for the Q_* value, we conclude that

$$Q_* = Q\frac{1 - \dfrac{t}{T}}{1 - \dfrac{t}{T}Q}. \tag{2.16}$$

Let, for example, the specified dependability of a navigational instrument be such that the predicted probability of its failure during the time $T = 5$ years is $Q = 0.01$. If the equipment was successfully operated during the time of $t = 4$ years, then the probability that it fails (say, because of materials aging) during the remaining year is

$$Q_* = 0.01 \frac{1 - 0.8}{1 - 0.8 \times 0.01} = 0.002016.$$

Thus, the probability of the failure-free operation of the instrument during this year is 0.998. This problem is revisited in Chapter 5.

7. Let an experiment be repeated many times, and on each occasion the occurrences or nonoccurrences of two events, A and B, are observed. Suppose that one only takes an interest in those outcomes for which the event B occurs, and disregards all the other experiments. The *conditional probability* of an event A, designated as $P(A/B)$, is defined as the probability of the event A, calculated *on the hypothesis* that the event B has occurred. With the concept of the conditional probability, the *multiplication rule of the probabilities* can be formulated as follows. The probability of the product of two events A and B is equal to the probability of one of them multiplied by the conditional probability of the other, provided that the first event has occurred:

$$P(AB) = P(A)P(B/A) = P(B)P(A/B). \tag{2.17}$$

In the case of three events, A, B, C, one can write this formula as

$$P(AH) = P(H)P(A/H) \tag{2.18}$$

and put first $H = BC$ and then apply the formula (2.17) for the probability of two events again. Then

$$P(ABC) = P(C)P(B/C)P(A/BC). \tag{2.19}$$

This rule can be generalized for an arbitrary number of events:

$$P(A_1 A_2 \ldots A_n) = P(A_1)P(A_2/A_1)P(A_3/A_1 A_2)\ldots P(A_n/A_1 A_2 \ldots A_{n-1}). \tag{2.20}$$

The formula (2.17) can be written as

$$P(A/B) = \frac{P(AB)}{P(B)}. \tag{2.21}$$

This relationship states that the conditional probability $P(A/B)$ of the event A on the hypothesis that the event B occurred is equal

Fundamentals of Applied Probability 15

to the probability $P(AB)$ of the product AB of the two events (both events occurred) divided by the probability $P(B)$ of the event B, which is assumed to be realized.

Two events A and B are *independent* if the occurrence of one of them does not affect the probability of the occurrence of the other:

$$P(A/B) = P(A), \quad P(B/A) = P(B). \tag{2.22}$$

For two independent events A and B, their simultaneous occurrence can be found as the product of their probabilities:

$$P(AB) = P(A)P(B). \tag{2.23}$$

For several independent events $A_1, A_2, \ldots, A_n$,

$$P(A_1, A_2, \ldots, A_n) = P(A_1)P(A_2)\ldots P(A_n). \tag{2.24}$$

8. Let $H_1, H_2, \ldots, H_n$ be a set of mutually exclusive events. One of these events, the event A, will definitely occur. In many cases, it might be easier to calculate the conditional probability $P(A/H_j)$, when the event A takes place in combination with another event H_j, than to calculate the probability $P(A)$ of the event A directly. Using the formula (2.18) and forming a sum with respect to the indices j, the following *total probability formula* can be obtained:

$$P(A) = \sum_{i=1}^{n} P(H_i)P(A/H_i). \tag{2.25}$$

9. If n independent trials are performed and in each of these trials an event A occurred with the probability p, then the probability P_m that this event will occur m times in a similar experiment is expressed by the following formula known as the *binomial distribution*:

$$P_m = C(n,m)p^m(1-p)^{n-m}, \tag{2.26}$$

where

$$C(n, m) = \frac{n!}{(n-m)! \, m!} \tag{2.27}$$

is the combination consisting of m different elements that can be rearranged by permuting them in $m!$ different ways. The binomial distribution is one of the most frequently used discrete random values distributions. It is addressed in greater detail later in this chapter.

Example 2.4. One hundred candidates ($n = 100$) have been tested to determine their HCF level. Ten of them passed the test, so that the event A (passing the test) has occurred with the probability $p = 0.1$. If only 10 candidates are tested ($m = 10$), how many of them are likely to pass the tests?

The formulas (2.26) and (2.27) yield the following:

$$C(n,m) = \frac{n!}{(n-m)!\ m!} = \frac{100!}{90!10!} = 1.7310 \times 10^{13};$$

$$P_m = C(n,m) p^m (1-p)^{n-m} = 1.7310 \times 10^{13} \times 0.1^{10} \times 0.9^{90} = 0.1319.$$

Hence, if 10 candidates are tested, only one of them will most likely pass the test. This is not surprising, but note that the predicted probability of passing the test is somewhat higher than 10%.

Example 2.5. A sinking vessel sends a rescue signal m times. The probability that this signal is received by n other vessels in the region is $P = \dfrac{1}{n}$. What is the probability that the given vessel that has a helicopter on her deck will be the one that will receive the rescue signal first?

The random variable x (the number of vessels that accept the rescue signal) can assume the following values: 0, 1, 2,...,n. The probability $P(n,k)$ that this variable will assume the value k (i.e., the probability that the given vessel will receive the rescue signal k times) can be determined by using the binomial distribution formula (2.26):

$$P(n,k) = C(m,k) p^k (1-p)^{m-k} = \frac{m!}{k!(m-k)!}\left(\frac{1}{n}\right)^k\left(1-\frac{1}{n}\right)^{m-k}. \quad (2.28)$$

For $k = 1$, this formula yields

$$P = C(m,1)\frac{1}{n}\left(1-\frac{1}{n}\right)^{m-1} = \frac{m}{n}\left(1-\frac{1}{n}\right)^{m-1}. \quad (2.29)$$

If, for example, the sinking vessel sent a rescue signal $m = 100$ times and there are $n = 50$ vessels in the region, then the probability that the given vessel will receive the rescue signal is

$$C(n,m_*) p^{m_*} (1-p)^{n-m_*} \geq C(n,m_*-1) p^{m_*-1} (1-p)^{n-m_*+1}$$

$$P = \frac{100}{50}\left(1-\frac{1}{50}\right)^{100-1} = 0.2707.$$

Fundamentals of Applied Probability 17

Example 2.6. The aerospace shuttle flight takes 50 days. What is the probability that an off-normal solar radiation that happens once in 100 days will occur during this flight?

The probability of such a radiation is $p = 0.01$. The result obtained in the previous example enables calculation of the sought probability as follows:

$$P_1 = C(50.1)(0.01)^1 (0.99)^{49} = 50 \times 0.01 \times 0.6111 = 0.3056.$$

Example 2.7. A human undertakes n consecutive actions. The probability p of the event that he/she makes an error is the same in every action. What is the most probable sequential number m_* when such an error will occur?

Although the answer to this question can be obtained assuming that the events of interest follow the binomial distribution, in this example, simple logic is used to solve the problem. In the most general case $0 \prec m_* \prec n$, the following two conditions

$$C(n, m_*) p^{m_*} (1-p)^{n-m_*} \geq C(n, m_* + 1) p^{m_* + 1} (1-p)^{n-m_* - 1}$$
$$C(n, m_*) p^{m_*} (1-p)^{n-m_*} \geq C(n, m_* - 1) p^{m_* - 1} (1-p)^{n-m_* + 1}$$

(2.30)

must be fulfilled. These conditions are equivalent to simpler conditions

$$(m_* + 1)(1 - P) \geq (n - m_*) p; \quad (n - m_* + 1) p \geq m_* (1 - p). \quad (2.31)$$

As follows from these inequalities, the sought most probable number m_* of an erroneous action must be an integer that satisfies the condition

$$(n+1)p - 1 \leq m_* \leq (n+1)p. \quad (2.32)$$

The solution is provided by the integer m_* that is the closest to the product np. Indeed, when np is an integer, and $m_* = np$, the inequality (2.32) yields $0 \leq p \leq 1$, as it is supposed to be for a probability p being a random variable. As a matter of fact, $\bar{x} = np$ is the mean value of the binomial distribution (see Section 2.3).

10. If, before an experiment is conducted, the probabilities $P(H_1), P(H_2), \ldots, P(H_n)$ of the hypotheses $H_1, H_2, \ldots, H_n$ were evaluated, and the experiment resulted in an event A, then the new (conditional) probabilities of the hypotheses can be evaluated by the *Bayes formula:*

$$P(H_i/A) = \frac{P(H_i)P(A/H_i)}{\sum_{i=1}^{n} P(H_i)P(A/H_i)}, \quad i = 1, 2, \ldots, n. \quad (2.33)$$

This formula, widely used in numerous applications, can be obtained by introducing the formula (2.18) and the formula (2.25) into the formula (2.21), in which the event B is replaced with the event H. In the formula (2.33), $p(H_1), P(H_2),\ldots,P(H_n)$ are *prior probabilities* and $P(H_1/A), P(H_2/A),\ldots,P(H_n/A)$ are *posterior or inverse probabilities*. Bayes formula (2.33) makes it possible to "revise" the probabilities of the initial hypotheses based on the new experimental data. Bayes formula indicates in a formal and in an organized fashion how new information can be used to update prior knowledge about the probabilities of random events of interest.

Example 2.8. The results of n medical analyses taken from a patient are fed into a diagnostics apparatus. Each analysis may be erroneous with the same probability p. The probability $P = P(m)$ of correct diagnosis is certainly a function of the number m of correct analyses. Diagnoses for k patients were made. What is the probability P_* that at least one erroneous diagnosis is made?

The hypotheses concerning the number of correct analyses made are: $H_0 = \{$not a single correct analysis$\}$; $H_1 = \{$exactly one correct analysis$\}$;...; $H_m = \{$exactly m correct analyses$\}$;...; $H_n = \{$exactly n correct analyses$\}$. The probability of the event H_m for any m can be calculated in this example by the formula (2.26) of the binomial distribution. This yields

$$P(H_0) = p^n; P(H_1) = n(1-p)p^{n-1};\ldots;P(H_m) = C(n,m)(1-p)^n p^{n-m};\ldots;$$

$$P(H_n) = (1-p)^n. \tag{2.34}$$

Using the total probability formula (2.25), the probability that an erroneous diagnosis was made for a certain patient is

$$P_1 = \sum_{m=0}^{n} C(n,m)(1-p)^m p^{n-m}. \tag{2.35}$$

The probability that no erroneous diagnosis was made for this patient is $1 - P_1$, and the probability that none of the k diagnoses was erroneous for this patient is $(1-P_1)^k$. The probability that at least one erroneous diagnosis was made is therefore

$$P_* = 1 - (1-P_1)^k. \tag{2.36}$$

Example 2.9. A manager has to hire the best performer out of 10 candidates. If he decides not to hire a certain candidate, the candidate will leave and will never come back. How should the right hiring strategy be established?

Fundamentals of Applied Probability   19

This problem is known as the *choosy bride problem*. A choosy bride could accept a proposal, and, if she does so, the process of selecting the best match is over. If she rejects the proposal, the groom is irretrievably lost. The bride follows a simple strategy: not to accept a proposal from a groom who is, by her standards, worse than his predecessors. Obviously, there is a risk involved each time, when the proposal is made: the bride can reject the absolutely best proposal in the hope of receiving a better one later. The question that the bride (the manager) asks herself (himself) is: what is the probability that the k-th groom (candidate) is the best out of the entire "set" of m grooms (candidates), both those having been rejected and the future ones? Actually, this is an example of a decision-making problem, in which one has to determine if the extra information justifies the expense of obtaining it. The optimal strategy is that the bride (manager) should turn down the proposals of the first k bridegrooms (candidates) and accept the proposal of the first bridegroom (candidate) who is better than the previous ones. The number k should be determined from the double inequality (no derivation is provided here):

$$\frac{1}{k+1}+\frac{1}{k+2}+\cdots+\frac{1}{m-1} \leq 1 \prec \frac{1}{k}+\frac{1}{k+1}+\cdots+\frac{1}{m-1} \qquad (2.37)$$

With such a strategy, the probability that the best bridegroom (candidate) is chosen out of m candidates is

$$p = \frac{k}{m}\left(\frac{1}{k}+\frac{1}{k+1}+\cdots+\frac{1}{m-1}\right). \qquad (2.38)$$

If the bride (manager) has to choose the best match (candidate) out of 10 bridegrooms (candidates), then the inequality (2.37) yields: $k=3$. Hence, the choosy bride should reject the proposals of the first three bridegrooms and then accept the proposal of the first candidate who, by her standards, turns out better than his predecessors. So, the probability that the bride (hiring manager) will select, using the above strategy, the best party (candidate) is

$$p = \frac{3}{10}\left(\frac{1}{3}+\frac{1}{4}+\cdots+\frac{1}{9}\right) = 0.399.$$

If the number m is large ($m \to \infty$), this probability can be found as $p = \frac{1}{e} = 0.368$. One encounters, in a way, a similar situation when choosing the most suitable astronaut based on testing candidates on a flight simulator or otherwise.

Example 2.10. In the previous problem assess the probability that no better choices will appear after the k-th one using the concept of conditional probability. In effect, one would like to find out how the occurrence of the event A (the k-th object is the best out of the total m objects) is affected by the event B (the last out of k examined objects is the absolute best).

The probability $P(AB)$ of the event AB (both events A and B occur) is $P(AB) = P(B)P(A/B)$, where $P(B)$ is the probability of occurrence of the event B and $P(A/B)$ is the conditional probability of the occurrence of the event A provided that the event B took place. The bride (manager) is interested in this latter probability.

Let us determine the probabilities $P(AB)$ and $P(B)$. Since the event B contains the event A (if the object is the absolute best, it will be the best also among the first k objects), the event AB coincides with the event A. As to the probability of the event B, it can be calculated as the probability that the best out of k objects will be found at the fixed k-th location as a result of a random permutation of k different objects. In accordance with the procedures of counting sample points, the probability $P(B)$ can be evaluated as

$$P(B) = \frac{(k-1)!}{k!} = \frac{1}{k}. \qquad (2.39)$$

Here $k!$ is the number of permutations for k objects and $(k-1)!$ is the number of permutations of $k-1$ objects, compatible with the condition that the best object is fixed at the k-th location. The probability $P(A)$ can be obtained in a similar way, by calculating the probability that as a result of the random permutation of m different objects, the particular, the best, object will occupy the k-th location:

$$P(A) = \frac{(m-1)!}{m!} = \frac{1}{m}. \qquad (2.40)$$

The sought probability that no better choices will appear after the k-th one is

$$P(A/B) = \frac{P(AB)}{P(B)} = \frac{P(A)}{P(B)} = \frac{k}{m}. \qquad (2.41)$$

This result indicates particularly that if the bride (manager) can wait, she/he should do so, since the probability to find the best match (employee) increases with an increase in the number k. If, for example, as in the previous example, $m = 10$ and $k = 3$ the sought probability is $P(A/B) = 0.3$ (compare with the results obtained in Example 2.9).

Example 2.11. The probability that the first weather forecaster predicts the flight weather correctly is p_1, and the probability that the second forecaster predicts the flight weather correctly is p_2. The first forecaster predicted fair weather for the duration of the flight, and the second one predicted bad weather. What is the probability that the first forecaster is right?

Let A be the event "the first forecaster is right" and B be the event "the second forecaster is right," so that $P(A) = p_1$ and $P(B) = p_2$. Since the two forecasters made different predictions, the event $A\bar{B} + \bar{A}B$ took place

Fundamentals of Applied Probability 21

(either the first forecaster is right and the second one is wrong, or the first forecaster is wrong and the second forecaster is right). The probability of this event can be evaluated as

$$P(A\bar{B}+\bar{A}B)=P(A\bar{B})+P(\bar{A}B)=P(A)P(\bar{B})+P(\bar{A})P(B)=p_1(1-p_2)+(1-p_1)P_2.$$
(2.42)

The weather will be fair if the event $A\bar{B}$ (the first forecaster is right and the second one is wrong) takes place. The probability of this event is

$$P(A\bar{B})=\frac{p_1(1-p_2)}{p_1(1-p_2)+(1-p_1)p_2}.$$
(2.43)

If, for example, $P_1 = P_2 = 0.5$, then $P(A\bar{B})=0.5$ as well, but if $p_2 = 0.5$ and $p_1 = 0.2$, then the probability that the first forecaster is right is

$$P(A\bar{B})=\frac{0.5\times0.8}{0.4+0.1}=0.8.$$

Example 2.12. At the given segment of a spacecraft route the forecasted environmental conditions (such as, for instance, solar radiation) are such that 2 out of 60 days in the given area of the outer space might be dangerous to the spacecraft and/or the astronauts. The spacecraft will be in the dangerous area of the outer space for about half a day. What is the probability that the spacecraft will encounter dangerous environmental conditions?

Since the spacecraft can be found in the potentially dangerous area for half a day, this time can be taken as a proper time unit. Then the problem can be reformulated as follows. The total number of the time units is $n = 120$, and $k = 4$ of them can be dangerous for the spacecraft and/or for its crew. The spacecraft travels in this region of the outer space for $n = 1$ unit of time. What is the probability that a dangerous situation happens during this time?

Examine first a more general and formally related problem. A batch of n articles contains k faulty ones. A sample of m articles is taken at random from the batch for inspection. What is the probability that this sample contains l faulty articles? The total number of ways by which one can take m articles out of n articles is $C(n, m)$. Favorable outcomes are those when l faulty articles are taken out of the total number k of faulty articles, while the remaining $m - l$ articles taken out of the total number of $n - k$ sound articles are sound ones. The l faulty articles can be taken out of the total number k of faulty articles in $C(k, l)$ ways, and $m - l$ articles can be taken out of the total number of $n - k$ of sound articles in $C(n - k, m - l)$ ways. The total number of favorable cases is therefore $C(k,l)$ $C(n - k, m - l)$, and the sought probability can be evaluated as

$$P=\frac{C(k,l)C(n-k,m-l)}{C(n,m)}.$$
(2.44)

With $n = 120$, $m = 1$, $l = 1$, $k = 4$, this formula yields:

$$P = \frac{C(k,l)C(n-k,m-l)}{C(n,m)} = \frac{k!(n-k)!(n-m)!m!}{(k-l)!l!(n-k-m+l)!(m-l)!n!}$$

$$= \frac{4!116!119!}{3!116!120!} = \frac{1}{30} = 0.0333.$$

Thus, the likelihood that the spacecraft encounters dangerous conditions is 3.33%.

2.3 Discrete Random Variables

2.3.1 Probability Characteristics

A *random variable* is a variable, which, as a result of an experiment with an unpredictable outcome, assumes a certain value that is unknown prior to the experiment. No matter how carefully a process is run, an experiment is performed, or a measurement is taken, there will be differences (variability) in its repeatability due to the inability of an actual system to control or to predict completely all possible influences. Examples are population of a town or a country, time to failure in a machine, time until a human makes an error, the level of the MWL or cognitive workload, HCF, fatigue lifetime of a material, stress level in a device and its remaining useful lifetime (RUL), current or voltage in an IC, light output of a photonic device, wind gust velocity, level of atmospheric turbulence, and so on. Random variables are usually denoted by capital letters from the end of the Latin alphabet, $X,Y,Z,...$, and the possible values of these variables (*realizations*)—by the corresponding small letters, x, y, $z,...$. A random event can be viewed as a special case of a random variable, namely, as a random variable, which assumes only two values: "unity," when this event takes place, and "zero," when it does not.

The most general form to define a random variable is to establish *the law of probability distribution* for this variable. Such a law is a rule that can be used to find the probability of the event related to the random variable of importance. It could be the probability that the variable assumes a certain value or falls within a certain interval. The most general form of a probability distribution law is a *distribution function*—the probability that a random variable X assumes a value that is smaller than the given (nonrandom) value x:

$$F(x) = P\{X \prec x\}. \tag{2.45}$$

Being a probability, this function possesses the following major properties: $F(-\infty) = 0$ and $F(-\infty) = 1$, so that $0 \le F(x) \le 1$. The function $F(x)$ always

Fundamentals of Applied Probability 23

increases with an increase in x. A random variable is *discrete* if it has a finite or countable set of possible values that can be enumerated. A discrete random variable X is defined if, for every value x_k, $k = 1,2,...$, of this variable, a corresponding probability p_k is known (given). This probability determines the objective likelihood with which this value can be realized. The simplest distribution for a discrete random variable X is an *ordered sample*, an *ordered series*, or a *frequency* table. This is a table whose top row contains the values x_k, $k = 1,2,...$, of the random variable in the ascending order, and the bottom row contains the corresponding probabilities p_k, $k = 1,2,...$:

X:	x_1	x_2	...	x_k	...
	P_1	P_2	...	P_k	...

In this table, P_k is the probability that the random variable X assumes the value x_k:

$$P_k = P\{X = x_k\}. \tag{2.46}$$

Clearly, the sum of all the probabilities P_k should be equal to "unity" (*condition of normalization*):

$$\sum_i P_i = 1. \tag{2.47}$$

The *mean (value)*, or the *mathematical expectation*, of a discrete random variable X is the sum of the products of all its values x_k and their probabilities P_k:

$$E(X) = \bar{x} = \prec x \succ = \sum_k x_k p_k. \tag{2.48}$$

The sample mean can be simply computed as the sum of the observed values of the random variable x_k divided by the number of observations:

$$E(X) = x = \prec x \succ = \frac{1}{n} \sum_{k=1}^{n} x_k. \tag{2.49}$$

A *centered random variable* is the difference between the random variable X and its mean:

$$\overset{0}{X} = X - \bar{x}. \tag{2.50}$$

The *variance* of a random variable X is the mean value of the square of the corresponding centered random variables:

$$D(X) = D_x = \text{var}(X) = \sum_k (x_k - \bar{x})^2 p_k. \tag{2.51}$$

The sample variance can be calculated by the formula

$$D_x = \frac{1}{n-1} \sum_{k=1}^{n} (x_k - \bar{x})^2. \tag{2.52}$$

The $n - 1$ term, not the n term, in the denominator of this formula occurs, because the statistical theory shows that by dividing by $n - 1$, instead of n, gives a better estimate of the population variance. This is because a sample variance always underestimates the population variance, since the value of the sample mean is not usually the same as the value of the population mean. Because of that, the sum of the deviations of the population values around the sample mean is usually smaller than the sum of the squares of the deviations about the population mean.

The *mean square deviation*, or the *standard deviation*, or, simply, *the standard*, of a random variable X is the square root of its variance:

$$\sigma_x = \sqrt{D_x}. \tag{2.53}$$

Unlike the variance, the standard deviation is measured in the same units as the random variable itself, and provides, therefore, more convenient information of the variability of the random value of interest.

In addition to the major probabilistic characteristics $\bar{x}$ and D_x (or σ_x), it is often useful to have measures of symmetry of the distribution of the random variable about the center and of how *peaked* the data are over the central region. Those characteristics are called *skewness* and *kurtosis*, respectively, and are defined as the expected values of the third and the fourth moments of the given variable about its mean:

$$\mu_3 = E\left[(x - \bar{x})^3\right] = \prec (x - \bar{x})^3 \succ, \quad \mu_4 = E\left[(x - \bar{x})^4\right] = \prec (x - \bar{x})^4 \succ. \tag{2.54}$$

A *unimodal* (single-peak) distribution with an extended right "tail" has a positive skewness and is referred to as *skewed right*. *Skewed left* implies a negative skewness and a corresponding extended left tail. A symmetric distribution has a zero skewness. As to the kurtosis, it characterizes the relative flatness (peakedness) of the distribution and indicates how "heavy" its tails are. Sometimes the widespread *normal (Gaussian) distribution* (see "Continuous Random Variables" section later in this chapter) is used as a suitable measure of reference for the kurtosis of the given distribution. The kurtosis of the normal distribution is $\mu_4 = 3$. If in the given distribution, weather discrete or continuous, the kurtosis is below $\mu_4 = 3$, the distribution is *platykurtic* (mild peak); if the kurtosis' fourth moment is above $\mu_4 = 3$, the given distribution is *leptokurtic* (sharp peak). Sometimes the magnitudes of the skewness and kurtosis, along with the *goodness-of-fit criteria* (see "Reliability" section), make it possible to check the validity of the assumed (and based on statistical data) probability distribution for a particular random variable of interest.

Fundamentals of Applied Probability

Example 2.13. The probability that a navigator eventually makes an error as a result of interpreting repetitive computer (navigation instrumentation) information, checking this information many times and getting tired is equal to p. The discrete random variable of interest in such a situation is the sequential number of checking and interpreting this information. What is the distribution of this random variable?

The probability that the navigator will not make an error after the k-th checking and interpreting the obtained information can be found as a product of the probability that he/she will not make an error after the $k - 1$ action and will make an error as a result of his/her k-th action. This probability is $(1-p)^{k-1} p$. Hence, the possible values (realizations) 1, 2, 3,...,k... of the random variables in question have the probabilities $p, (1-p)p, ..., (1-p)^{k-1} p, ...$. The probability that the navigator will not make an error after his/her k-th action is $(1-p)^k$, and the probability distribution function of such a sequence of actions is

$$F(k) = 1 - (1-p)^k. \tag{2.55}$$

If, for example, $p = 0.01$ and $k = 100$, then $F(100) = 1 - 0.99^{100} = 0.634$.

Let us make sure that the normalization requirement (2.47) is fulfilled. Using the formula for the sum of an infinite geometric progression, we have

$$\sum_{k=1}^{\infty} (1-p)^{k-1} p = \frac{p}{1-p} \sum_{k=1}^{\infty} (1-p)^k = \frac{p}{1-p} \frac{1-p}{p} = 1. \tag{2.56}$$

In effect, the discrete process addressed in this example is a rather general binomial distribution (2.26) that characterizes the probability of the sequence of random events.

This distribution deals with a sequence of independent tests (*Bernoulli trials*). For the binomial distribution to be applicable, the outcome of each trial must be independent of all the other trials, only one of the two mutually exclusive outcomes ("success" or "failure") is possible, the probability of occurrence of "success" and "failure" in each trial should be the same for all trials, and the order in which the "successes" and "failures" occur is not important. The mean, the variance, and the coefficient of variation of the binomial distribution are

$$\bar{x} = np, \quad D_x = np(1-p), \quad \text{cov} = \frac{\sqrt{D_x}}{\bar{x}} = \sqrt{\frac{1-p}{np}}. \tag{2.57}$$

See also Example 2.7.

Here are several other well-known situations, where the binomial distribution is used:

- *Urn problem*: An urn contains white and black balls. The probability of taking out a white ball (event A) is $P(A) = p$. Then the probability of taking out a black ball (event $\bar{A}$) is $P(\bar{A}) = 1 - p$.

One carries out n trials and, to make the trials independent, each time puts the drawn ball back. What is the probability P_m that the white ball will appear m times during the trial?

- *Tossing of a coin:* A coin is tossed n times. The probability of the two possible outcomes ("heads" or "tails") is the same and is equal to $\frac{1}{2}$. What is the probability that the "heads" will come up m times after n tossings?
- *Newborn babies:* What is the probability that m out of n newborn babies are boys, if the probability of the event "the newborn baby is a boy" is $p = 0.51$?
- *Random motion:* A Brownian particle moves one unit to the right with the probability p and one unit to the left with the probability $1 - p$. What is the probability that the particle will make m moves to the right if the total number of moves is equal to n?

2.3.2 Poisson Distribution

The probability of occurrence of a discrete random variable X that follows *Poisson distribution* can be calculated by the following formula:

$$P_m = P\{X = m\} = \frac{a^m}{m!}e^{-\alpha}, \quad \alpha \succ 0, m = 0, 1, 2 \ldots \quad (2.58)$$

The mean and the variance of the Poisson distribution are the same:

$$\bar{x} = \alpha, \quad D_x = \alpha. \quad (2.59)$$

The Poisson distribution can be obtained from binomial distribution by putting $p \to 0$ and $n \to \infty$, given that $\alpha = np = const$. The Poisson distribution is used to make approximations, when one deals with a large number of independent trials, in each of which the event of interest is rare (i.e., whose probability of occurrence is small). Examples include earthquakes, floods, accidents, telephone calls, human errors, and extraordinarily high levels of MWL. The Poisson distribution can be used, for example, to estimate, on the probabilistic basis, the number of meteorites falling in the given region of space, especially when the location of the points in space is random and their number is small.

A one-dimensional Poisson's distribution is known as *flow of events*, or *traffic*. It is a sequence of the train of homogeneous events occurring at random points in time. The average number of events per unit time is the *intensity* of the flow. The flow can be either constant or time dependent. The flow of events is said to be *without aftereffects*, if the probability of the number of events that can be found within the given time interval is independent of the number of events in any other nonoverlapping interval. A flow of events is *ordinary*, if the probability of the occurrence of two or more events that can be found within an elementary time interval Δt is negligibly small compared

Fundamentals of Applied Probability 27

to the probability of occurrence of just one event. An ordinary flow of events without an aftereffect is *Poisson flow*, or *Poisson traffic*. If certain events form such a flow, then the number X of events occurring within an arbitrary time interval $(t_0, t_0 + \tau)$ has a Poisson distribution, in which the parameter

$$\alpha = \int_{t_0}^{t_0 + \tau} \lambda(t)dt \qquad (2.60)$$

is the mean value of the number of points in this interval, and $\lambda(t)$ is the *intensity of the flow*. If the intensity of the flow is time independent, the flow is said to be *steady-state* or *elementary* flow of the Poisson type. For such a flow $\sigma = \lambda t$. When dealing with problems associated with the Poisson distribution, it is convenient to use tabulated values of the functions:

$$P(m,\alpha) = \frac{\alpha^m}{m!}e^{-\alpha}, \quad R(m,\alpha) = \sum_{k=0}^{m} \frac{a^k}{k!}e^{-\alpha}. \qquad (2.61)$$

The function $R(m,\alpha)$ is the probability that the random variable X, which follows the Poisson distribution, assumes the value α not exceeding m: $R(m,\alpha) = P\{X \le m\}$.

Example 2.14. The probability of human error during the flight/mission is estimated to be $p = 10^{-4}$. What is the probability that two out of 1000 navigators will make an error during their flights, assuming that the flight conditions, the MWLs, and HCFs of all of the navigators are the same, and that Poisson's distribution is applicable for the rare events of the navigators' possible errors?

The parameter $\alpha = np = 1000 \times 10^{-4} = 0.1$. The formula (2.58) with $m = 2$ yields

$$P_2 = \frac{0.1^2}{2!}e^{-0.1} = 0.004524.$$

If the more general formula based in the binomial distribution (2.26) were used (this distribution is does not require that the events are rare and the probability p is small), then, with $n = 1000, m = 2, p = 10^{-4}$, so that

$$C(n,m) = \frac{n!}{(n-m)!m!} = \frac{1000!}{998!2} = \frac{999 \times 1000}{2} = 499500,$$

we have

$$P_m = C(n,m)p^m(1-p)^{n-m} = 499500 \times 10^{-8} \times (1-10^{-4})998 = 0.004520.$$

The results are close.

Example 2.15. What is the probability that in the previous example there will be not more than two human errors? Using the second formula in (2.61), we find

$$R(2,0.1) = \sum_{k=0}^{2} \frac{0.1^k}{k!} e^{-0.1} = e^{-0.1}\left(1 + 0.1 + \frac{0.01}{2}\right) = 0.9998.$$

Example 2.16. It has been observed that the average number of cars making a left turn at the given intersection, while the green left turn signal is on, is equal to six. The arrival of cars to the traffic light is a random and a rare event that follows Poisson distribution. The designated left-hand turn lane is suggested to be effective 95% of the time. How many cars should this lane be able to accommodate?

The second formula in (2.61) yields for $\alpha = 6$

$$R(m,6) = e^{-6} \sum_{k=0}^{m} \frac{6^k}{k!}.$$

The calculations are carried out in Table 2.1. As evident from the calculated data, a lane should be able to accommodate not less than 10 cars in order to meet the requirement that 6 cars in average are served per cycle and be effective 95% of the time.

Example 2.17. The outer space particles hitting the surface of a spacecraft form a field with a density λ (particle per meter squared). The equipment, exposed to the "rain" of such particles, occupies the area S on the surface of the spacecraft. The equipment fails if two particles hit it. The probability that the equipment fails as a result of being hit with one particle is p. What is the probability that the equipment fails?

The mean number of particles hitting the surface S of the equipment is $\alpha = \lambda S$. Examine the probabilities of the following two hypotheses: $H_1 =$ {one particle hits the equipment}; and $H_2 =$ {not less than two particles hit the equipment}. Using Poisson distribution, we find that

$$P(H_1) = \lambda S e^{-\lambda S}; \quad P(H_2) = R_2 = 1 - P_0 - P_1 = \bar{R}(1, \lambda S) = 1 - e^{-\lambda S}(1 + \lambda S).$$

TABLE 2.1

Number of Cars in a Designated Left-Hand Turn Lane

k	0	1	2	3	4	5	6
$6^k/k!$	1	6	18	36	54	64.8	64.8
$R(m,6)$	0.00248	0.01735	0.06197	0.15120	0.28506	0.44568	0.60630
k	7	8	9	10	11	12	>12
$6^k/k!$	55.543	41.543	27.668	16.609	9.049	4.544	3.539
$R(m,6)$	0.74398	0.84695	0.91554	0.95670	0.97914	0.99040	0.9992

Fundamentals of Applied Probability

The conditional probabilities of the event A "the equipment fails" are $P(A/H_1) = p; P(A/H_1) = 1....$ Then the sought probability that the equipment fails if two particles hit it can be found as a complete probability as follows:

$$P(A) = \lambda Spe^{-\lambda S} + 1 - e^{-\lambda S}(1 + \lambda S) = 1 - e^{-\lambda S}\left[1 + \lambda S(1 - p)\right]. \qquad (2.62)$$

2.4 Continuous Random Variables

2.4.1 Probability Characteristics

A continuous (nondiscrete) random variable X is characterized by an uncountable set of possible values. The continuous random variable is defined if its cumulative probability distribution function

$$F(x) = P\{X \prec x\} \qquad (2.63)$$

is known. This function can be interpreted as the fraction of values in the population less than or equal to x. This function is continuous for any x, and so is its derivative (the probability density function)

$$f(x) = \frac{dF(x)}{dx}, \qquad (2.64)$$

except, perhaps, at some particular points. This function can be interpreted as the fraction of the population values that can be found in the interval dx. The probability density distribution function (frequency of the outcomes) can be obtained from the experimental histogram of frequencies. The probability of a particular individual value of a continuous random variable is equal to zero. The probability that the random variable X can be found within the values $x = a$ and $x = b$ can be evaluated as

$$P\{a \leq X \leq b\} = F(b) - F(a). \qquad (2.65)$$

As follows from (2.64), the cumulative probability distribution function is

$$F(x) = \int_{-\infty}^{x} f(x)dx. \qquad (2.66)$$

Since $F(\infty) = 1$, then

$$\int_{-\infty}^{\infty} f(x)dx = 1. \qquad (2.67)$$

This formula expresses the condition of normalization for a continuous random variable (compare with the condition [2.47] for a discrete random variable). The probability that a continuous random variable X falls within an interval from a to b can be calculated as

$$P\{a \leq X \leq b\} = \int_a^b f(x)dx = F(b) - F(a).$$

(2.68)

The mean value (mathematical expectation) and variance of a continuous random variable X are

$$\bar{x} = \int_{-\infty}^{\infty} xf(x)dx$$

(2.69)

$$D_x = \triangleleft (x - \bar{x})^2 \triangleright = \int_{-\infty}^{\infty} (x - \bar{x})^2 f(x)dx.$$

(2.70)

The square root of variance is *mean square deviation*, or *standard deviation*, or just *standard*. The standard deviation is often used to approximate the range of the possible values of a random variable. The *coefficient of variation*

$$\mathrm{cov}_x = v_x = \frac{\sigma_x}{\bar{x}}$$

(2.71)

is used as a characteristic of the "degree of randomness" of a random variable. It is equal to zero for a deterministic variable and is equal to infinity for an "absolutely uncertain" variable. In the probabilistic reliability theory a reverse (nonrandom) characteristic, called *safety factor*, is used, $SF_x = \dfrac{\bar{x}}{\sigma_x}$, and the mean and the standard deviation are calculated for the random safety margin $SM_x = C - D$, which is the difference between the random capacity ("strength") C and the random demand ("stress") D.

The k-th moment about the origin for a continuous random variable X is

$$m_k(x) = \triangleleft x^k \triangleright = \int_{-\infty}^{\infty} x^k f(x)dx.$$

(2.72)

The k-th central moment is

$$\mu_k(x) = \int_{-\infty}^{\infty} (x - \bar{x})^k f(x)dx.$$

(2.73)

The use of these moments is helpful for many applications.

Fundamentals of Applied Probability

Example 2.18. The probability p of encountering a molecule of a gas inside a small volume dv is proportional to the size of this volume and is, say, adv. The distance of this molecule from the nearest adjacent one is a random variable r. What is the probability distribution of this distance?

The following two conditions should be fulfilled: (1) there should be a molecule inside a spheric layer $4\pi r^2 dr$ and (2) the distance between the molecules should not be smaller than r. The probability of the first event is $p = \alpha(4\pi r^2 dr)$ and the sought probability of the second event is $1 - F(r)$ These two events are statistically independent and the probability of the sought event is

$$f(r)dr = 4\pi\alpha r^2 \left[1 - F(r)\right]dr. \tag{2.74}$$

Since $F'(r) = f(r)$, this equation can be written as

$$\frac{f'(r)}{r^2} - 2\frac{f(r)}{r} + 4\pi\alpha f(r) = 0. \tag{2.75}$$

This differential equation has the following solution:

$$f(r) = Cr^2 \exp\left(-\frac{4}{2}\pi\alpha r^3\right). \tag{2.76}$$

The constant C of integration can be determined from the condition of normalization (2.67) as follows: $C = 4\pi\alpha$. Then the solution (2.76) yields

$$f(r) = 4\pi\alpha r^2 \exp\left(-\frac{4}{3}\pi\alpha r^3\right). \tag{2.77}$$

This solution indicates that the probability density of the distance r between the molecules increases from zero to its maximum value, when this distance increases from zero to $r = (2\pi\alpha)^{-1/3}$ and then decreases to zero with the further increase in r.

2.4.2 Bayes Formula for Continuous Random Variables

If the probability of an event A depends on the value x assumed by a continuous random variable X, whose probability density function is $f(x)$, then the total probability of this event can be calculated on the basis of the formula:

$$P(A) = \int_{-\infty}^{\infty} P(A/x) f(x) dx. \tag{2.78}$$

Here

$$P(A/x) = P\{A/X = x\} \tag{2.79}$$

is the conditional probability of the event A on the hypothesis $\{X = x\}$ (i.e., the probability that is calculated provided that the random variable X assumed the realization x). If an event A occurs as a result of a certain experiment, and the probability of this event depends on a value assumed by a continuous random variable X, then the conditional probability density function of the random variable X, when the occurrence of the event A is taken into account, is

$$f_A(x) = f(x)\frac{P(A/x)}{P(A)}.$$
(2.80)

This relationship can be written, considering (2.78), as follows:

$$f_A(x) = \frac{f(x)P(A/x)}{\int\limits_{-\infty}^{\infty} P(A/x)f(x)dx}.$$
(2.81)

This is *Bayes formula* for a continuous random variable X. Bayes formula is simple, is easy to apply, and is used therefore in numerous applied science and engineering problems. Its shortcomings are the large volume of the required input information and strong suppression of the role of infrequent events.

Example 2.19. The blood pressure X of a human (tested for his/her eligibility to join the team of astronauts) is a continuous random variable with the probability density function $f(x)$. During the medical control of a group of such candidates, all those whose measured blood pressure was outside the normal interval (x_1, x_2) were rejected. Determine the probability density function for those candidates who met the $x_1 \leq x \leq x_2$ requirement.

As a result of testing it has been found that the probability of the event A "the candidate was not rejected" is

$$P(A) = \int\limits_{x_1}^{x_2} f(x)dx.$$
(2.82)

Since $P(A/x) = 0$ for $X = x \prec x_1$ and $X = x \succ x_2$, and $P(A/x) = 1$ for $X = x \in (x_1, x_2)$, the Bayes formula (2.81) yields:

$$f_A(x) = \frac{f(x)}{P(A)} = \frac{f(x)}{\int\limits_{x_1}^{x_2} f(x)dx}.$$
(2.83)

Fundamentals of Applied Probability 33

2.4.3 Bayes Formula as a Technical Diagnostics Tool

Bayes formula is widely used in medical diagnostics but can be effectively used also in technical diagnostics. Its objective is to recognize, in a continuous fashion and by using nondestructive means (i.e., without taking apart the object of interest), the typically hidden technical state (*health*) of the object (device) of interest and to assess the object's ability to continue its performance in the expected (specified) fashion. Technical diagnostics establishes the links between the observed (detected) signals, *symptoms of faults* (SoFs), and the "health" of the object. Technical diagnostics provides assistance in making a decision as to whether the device(s) or the instrumentation of interest are still sound or have become (still, hopefully, acceptably or) unacceptably faulty. There is always a risk that the interpretation of the obtained SoF signal might be a false alarm or, on the contrary, might lead to a "missing-the-target" decision. Technical diagnostics is supposed to assess the likelihoods of such possibilities. Technical diagnostics is focused on the most vulnerable elements (*weakest links*) of the design and can make use of the accelerated test data collected earlier, at the design stage.

This body of knowledge encompasses a broad spectrum of problems associated with obtaining, processing, and assessing diagnostics information, including diagnostics models, decision-making rules, and algorithms, and provides information for the subsequent prognostics and health monitoring effort. Technical diagnostics is supposed to come up with the appropriate solutions and recommendations, usually in the conditions of uncertainty and typically on the basis of limited information. The methods, techniques, and algorithms of technical diagnostics are naturally based on probabilistic risk analysis (i.e., are employed to analyze and quantify, on a probabilistic basis, the obtained diagnostic information). Application of Bayes formula in aerospace electronics enables assessment of the reliability of a particular malfunctioning device from the available general information for similar devices. It can be used also in a medical diagnostics effort, when there is a need to monitor the state of human health.

As has been indicated, the objective of the Bayes formula is to compare new information with what is known and considered typical for the system of interest and to use this information to revise the previous beliefs about the probabilities of the occurrence of the events of interest. Interpretation of the results of a diagnostic test, when Bayes formula is applied, depends not only on the current test result or results, but also on prior information that is "external" to this result. This information is combined with the new test data to obtain information of the true prevalence and the test characteristics. In this section, Bayes formula is interpreted in application to technical diagnostics problems in electronics and particularly in aerospace electronics.

Let an event S be the observed (detected) signal (SoF), such as, for example, measured elevated off-normal temperature, elevated leakage current, significant drop in the output light intensity, elevated amplitudes (power spectrum) of the induced vibrations, and so on, and the events D_i, $i = 1, 2, 3,...$ be detected deviations from the device's (instrumentations) normal operation conditions that might be responsible for the observed symptom or symptoms. It is assumed that one and only one of the device's elements is damaged to an extent that its detected off-normal performance has manifested itself in the observed symptom. Failures are rare events, and therefore, a simultaneous damage of two or more elements is deemed to be extremely unlikely and is excluded from consideration. It is assumed that based on the accumulated experience for the type of equipment of interest, one knows the typical probabilities $P(D_i)$ of failure of its particular elements. Then the problem can be formulated this way: the event (signal) S is observed for the given device in operation. What is the probability that it is the system's particular ith element that has become faulty and is therefore the root cause of the detected symptom?

Since the probability density function $f(x)$ in the Bayes formula (2.81) is a conditional probability, this formula can be written as

$$P(D_i/S) = \frac{P(D_i)P(S/D_i)}{\sum\limits_{j=1}^{n} P(D_j)P(S/D_j)}.$$ (2.84)

This formula determines the posterior probability $P(D_i/S)$ after the symptom S has been detected, from the prior probability $P(D_i)$ of the system's state, known from the previous experience. The formula (2.84) can be obtained from the complete probability formula

$$P(S) = \sum\limits_{j=1}^{n} P(D_j)P(S/D_j)$$ (2.85)

and the relationship

$$P(S)P(D_i/S) = P(D_i)P(S/D_i).$$ (2.86)

The formula (2.85) reflects an intuitively obvious postulate that if a system has several possible and incompatible ways to get transferred from the state D_j to the state S, the probability of such an event can be found as the sum of the conditional probabilities of occurrence of each of these possible and incompatible ways. The relationship (2.86) indicates that the probability of the simultaneous occurrence of the symptom S and the system condition (diagnosis) D_i can be obtained as the product of these probabilities. As follows from the formula (2.84),

$$\sum\limits_{i=1}^{n} P(D_i/S) = 1.$$ (2.87)

Fundamentals of Applied Probability   35

The formula (2.84) indicates, particularly, that the factor

$$\chi = \frac{1}{P(D_1) + P(D_2)\dfrac{P(S/D_2)}{P(S/D_1)}} \tag{2.88}$$

accounts for the change, based on the updated reliability information, in the initial probability that the device is still sound and, hence, its use could be continued with a high level of confidence.

> **Example 2.20.** It has been established, for example, from experience with the given types of devices subjected in actual operation conditions to elevated temperature and vibrations that 90% of the devices of interest do not typically fail during operation. It has been established also that the diagnostic symptom—an increase in temperature by 20°C above the normal (specified) level—is encountered in 5% of the devices. The technical diagnostics instrumentation has detected in a particular device the following two deviations ("symptoms of failure") from normal operation conditions: (1) increase in temperature by 20°C at the heat sink location (symptom S_1) and (2) increase in the vibration power spectrum by 20% (symptom S_2). These symptoms might be caused by the malfunction of the heat sink (state D_1) and/or of the vibration protection equipment (state D_2). From previous experience with similar devices and at similar operation conditions, it has been established that the symptom S_1 (increase in temperature) is not observed at normal operation condition (state D_3), and the symptom S_2 (increase in the power of the vibration spectrum) is observed in 5% of the cases (devices). It has been established also, based on accumulated experience with this type of devices, that 80% of them do not fail during the specified time of operation, 5% of the devices experience the malfunction of the heat sink (state D_1), and 15% of them are characterized by state D_2 (malfunction of the vibration protection equipment). Finally, it has been established that symptom S_1 (increase in temperature) is encountered in state D_1 (because of the malfunction of the heat sink) in 20% of the devices, and in state D_2 (because of the malfunction of the vibration protection system) in 40% of the devices; and that symptom S_2 (increase in the power of the output vibration spectrum) is encountered in state D_1 (malfunction of the heat sink) in 30% of the devices and in state D_2 (malfunction of the vibration protection system) in 50% of the devices. The above information can be summarized in the form of the diagnostics matrix shown in Table 2.2.

> **TABLE 2.2**
>
> Diagnostics Matrix
>
> | D_i | $P(S_1/D_i)$ | $P(S_2/D_i)$ | $P(S_3/D_i)$ |
> |-------|--------------|--------------|--------------|
> | D_1 | 0.20 | 0.30 | 0.05 |
> | D_2 | 0.40 | 0.50 | 0.15 |
> | D_s | 0.00 | 0.05 | 0.80 |

Thus, this matrix indicates that

1. The symptom S_1 (increase in temperature) is encountered in 20% of the cases because of the malfunctioning heat sink (state D_1), in 40% of the cases because of the malfunctioning vibration protection system (state D_2), and is never observed in normal operation conditions (state D_3).
2. The symptom S_2 (increase in the power of the vibration spectrum) is encountered in 30% of the cases because of the malfunctioning heat sink (state D_1), in 50% of the cases because of the malfunctioning vibration protection system (state D_2), and in 5% of the cases in normal operation conditions (state D_3).
3. The symptom S_3 (both heat transfer and vibration protection hardware work normally) is encountered in 5% of the cases because of the malfunctioning heat sink (state D_1), in 15% of the cases because of the malfunctioning vibration protection system (state D_2), and in 80% of the cases in normal operation conditions (state D_3).

Let us determine first that the device in which the 20°C increase in temperature has been detected is still sound. This can be done using the information that 90% of the devices of the type of interest do not typically fail during the designated time of operation and that the symptom S_1 (an increase in temperature by 20°C above the normal level) is encountered in 5% of these devices. The first message tells that the probabilities of the sound condition D_1 and the faulty condition D_2 in the general population of the devices under operation are $P(D_1) = 0.9$ and $P(D_2) = 0.1$, respectively. The second message tells that the conditional probabilities reflecting the actual situation with the given device are $P(S/D_1) = 0.05$ and $P(S/D_2) = 0.95$: only 5% of the devices function adequately, and 95% of them do not. The question asked is as follows: with this new information about a particular device, how did the expected probability $P(D_1) = 0.9$ that the device of interest is still sound change? In other words, how could one use the accumulated experience about the operational performance of the large population of this type of devices, considering the results of the actual field information for a particular device?

The Bayes formula yields the following:

$$P(D_1/S) = \frac{P(D_1)P(S/D_1)}{P(D_1)P(S/D_1) + P(D_2)P(S/D_2)}$$

$$= \frac{0.9 \times 0.05}{0.9 \times 0.05 + 0.10 \times 0.95} = 0.32$$

Thus, the probability that the device is still sound has decreased dramatically, from 0.90 for the typical (expected) situation

Fundamentals of Applied Probability

to as low as 0.32 because of the detected 20°C increase in the observed temperature and because such an increase is viewed as a failure of the device.

The factor χ that accounts for the change in the initial probability that the device is still sound and, hence, its use could be continued with a high level of confidence. The decrease in the probability of nonfailure would be much different, if only a slight decrease in the probability of nonfailure for the given device, based on the obtained symptom, is detected. Indeed, with $P(S/D_1) = 0.85$ (instead of 0.05) and $P(S/D_2) = 0.15$ (instead of 0.95), the factor χ would be as high as $\chi = 0.9808$, and the updated probability of nonfailure would be also high: $P(D_1/S) = 0.8827$.

Let us address now, using the information provided by Table 2.1, the performance of a device affected by the possible malfunction of the heat sink and/or the vibration protection system. The probabilities of the device states, when both symptoms, S_1 (faulty heat sink) and S_2 (inadequate vibration protection), have been detected, can be found using Bayes formula as follows:

$$P(D_1/S_1S_2) = \frac{0.05 \times 0.02 \times 0.30}{0.05 \times 0.20 \times 0.30 + 0.15 \times 0.40 \times 0.50 + 0.80 \times 0.00 \times 0.05}$$

$$= 0.09.$$

This is the probability that the device, for which both symptoms, malfunctioning heat sink and malfunctioning vibration protection system, have been detected, is in the state D_1 (i.e., failed because of the malfunctioning heat sink).

Similarly, one could find the probability $P(D_2/S_1S_2) = 0.91$ that the device is in the state D_2 (i.e., failed because of the malfunctioning vibration protection system). Since the device has failed, it cannot be in the nonfailure state D_3, and therefore, the probability that the device is still sound, despite the detected malfunctions of the heat sink and the vibration protection system, is zero: $P(D_3/S_1S_2) = 0$.

Let us determine the probability of the device's state if the measurements have indicated that there was no increase in temperature (the symptom S_1 did not take place), but the symptom S_2 (increase in the power spectrum of the induced vibrations) was detected. The absence of the symptom S_1 means that the symptom $\bar{S}_1$ of the opposite event took place, so that $P(\bar{S}_1/D_i) = 1 - P(S_1/D_i)$. Changing the probability $P(S_1/D_i)$ in the above diagnostics matrix for $P(\bar{S}_1/D_i)$, we find the following probability of the state D_1 of the device (the device failed because of the malfunctioned heat sink):

$$P(D_1/\bar{S}_1S_2) = \frac{0.05 \times 0.80 \times 0.30}{0.05 \times 0.80 \times 0.30 + 0.15 \times 0.60 \times 0.50 + 0.80 \times 1.00 \times 0.05}$$

$$= 0.12.$$

Similarly, we obtain

$$P(D_2/\bar{S}_1S_2)=0.46; \quad P(D_3/\bar{S}_1S_2)=0.41.$$

Determine now the probabilities of the device states when none of the symptoms took place. By analogy with the above calculations, we find

$$P(D_1/\bar{S}_1S_2)=\frac{0.05\times0.80\times0.70}{0.05\times0.80\times0.70+0.15\times0.60\times0.50+0.80\times1.00\times0.15}$$

$$=0.03.$$

Similarly, we have

$$P(D_2/\bar{S}_1\bar{S}_2)=0.05; \quad P(D_3/\bar{S}_1\bar{S}_2)=0.92.$$

Thus, when both symptoms, S_1 and S_2, are observed, the state D_1 (failure occurred because the heat sink is malfunctioning) has the probability of occurrence of 0.91. When none of these symptoms are observed, the normal state, D_3, is characterized by the probability 0.92 and, hence, is somewhat more likely to occur than the state, when both symptoms, S_1 and S_2, are observed. When the symptom S_1 (elevated temperature) is not observed, while the symptom S_2 (elevated vibrations) is, the probabilities of the states S_2 (vibration protection system is not working properly) and S_3 (both heat transfer and vibration protection hardware work normally) are 0.46 and 0.41, respectively. One could either accept this information and act accordingly (i.e., go ahead with a conclusion that it is the elevated temperature and not the elevated vibration that should be taken care of) or, since these probabilities are close, one might decide to seek additional information. Such information could be based on additional measurements and observations and/or should use other sources to obtain more accurate and more convincing diagnostics information (e.g., by modeling).

2.4.4 Uniform Distribution

A random variable X has a uniform distribution on the interval from a to b, if its probability distribution density on this interval is constant:

$$f(x)=\frac{1}{b-a} \quad \text{for } x \in (a,b) \text{ and } f(x)=0 \quad \text{for } x \notin (a,b). \tag{2.89}$$

The cumulative distribution function of the uniform distribution is

$$F(x)=\frac{x-a}{b-a}. \tag{2.90}$$

Fundamentals of Applied Probability 39

The mean value, the variance, and the coefficient of variation of this distribution are

$$\bar{x} = \frac{a+b}{2}, \quad D_x = \frac{(b-a)^2}{12}, \quad v_x = \frac{1}{\sqrt{3}}\frac{b-a}{b+a}. \tag{2.91}$$

The uniform distribution has the largest *entropy* (uncertainty) of all the possible distributions on the given finite interval (a, b). The term "entropy" was introduced and coined by Clausius in 1850 in connection with the second law of thermodynamics, meaning "transformation" (say, transformation of mechanical work into heat). This term can be applied, however, to evaluate the degree and state of uncertainty in any experiment, communication line, system, or situation. The notion of entropy will be used in several subsequent chapters to explain and to substantiate the physics underlying some probability distributions of human factor characteristics.

> **Example 2.21.** In Example 2.19, let the blood pressure be a uniformly distributed random variable. Then the denominator in the formula (2.81) is equal to one, and the conditional probability density simply coincides with the probability density function $f(x)$.

> **Example 2.22.** The subway trains are run with 2-minute time intervals. A passenger arrived at the subway station at a random moment of time. What is the probability that he/she will have to wait for a train not longer than half a minute?
> The probability density function for the random time T that the passenger will be waiting for a train is $f(t) = \frac{1}{2}, 0 \prec t \prec 2$.
> The sought probability is

$$P(T \prec 1/2) = \int_{0}^{1/2} f(t)dt = \frac{1}{4}.$$

2.4.5 Exponential Distribution

A random variable X has an *exponential distribution*, if its probability density function is

$$f(x) = \lambda e^{-\lambda x} \quad \text{for } x \succ 0 \text{ and } f(x) = 0 \quad \text{for } x \prec 0. \tag{2.92}$$

The mean value, the variance, the standard deviation, and the coefficient of variation of this distribution are

$$\bar{x} = \sigma_x = \frac{1}{\lambda}, \quad D_x = \frac{1}{\lambda^2}, \quad v_x = 1. \tag{2.93}$$

The exponential distribution has the largest entropy (uncertainty) of all the possible distributions with the same mean value.

> **Example 2.23.** When a human is operating, his/her error may occur at random moments of time. The random time T of operation until the first error occurs has an exponential distribution with parameter λ:
>
> $$f(t) = \lambda e^{-\lambda t}. \tag{2.94}$$
>
> When the error occurs, the necessary corrections are made within the time t_0, after which the operations continue. Find the probability that the time interval Z between successive errors exceeds $2t_0$. The time Z between successive errors is related to the time T until the first error occurs as $Z = T + t_0$. The distribution of the random variable Z is $f(z) = \lambda e^{-\lambda(z-t_0)}$ for $z \succ t_0$ and $f(z) = 0$ for $z \succ t_0$. The sought probability can be found as
>
> $$P\{Z \succ 2t_0\} = \int_{2t_0}^{\infty} f(z)dz = e^{-\lambda t_0}. \tag{2.95}$$

2.4.6 Normal (Gaussian) Distribution

The probability density of this distribution is

$$f(x) = \frac{1}{\sqrt{2\pi D_x}} \exp\left[-\frac{(x - \bar{x})^2}{2D_x} \right], \quad -\infty \le x \le \infty. \tag{2.96}$$

This distribution is defined by two parameters: the mean value $\bar{x}$ and the variance D_x of the random variable X. The normal distribution is "bell shaped" (i.e., symmetric with respect to the vertical line $x = \bar{x}$). The maximum (most likely) value of the distribution takes place for $x = \bar{x}$ and is

$$f_{max} = \frac{1}{\sqrt{2\pi D_x}}. \tag{2.97}$$

The cumulative distribution function of the normal distribution is

$$F(x) = \int_{-\infty}^{x} f(x)dx = \frac{1}{2} + \int_{0}^{x} f(x)dx = \frac{1}{2}\left[1 + \Phi(\alpha)\right], \tag{2.98}$$

where

$$\Phi(\alpha) = erf(\alpha) = \frac{2}{\sqrt{\pi}} \int_{0}^{\alpha} e^{-t^2}\, dt, \quad \alpha = \frac{x - \bar{x}}{\sqrt{2D_x}}, \tag{2.99}$$

Fundamentals of Applied Probability 41

is *Laplace function,* or *the probability integral,* or the *error function.* This function has the following major properties:

$$\Phi(0) = 0; \quad \Phi(-\alpha) = -\Phi(\alpha); \quad \Phi(\infty) = 1. \tag{2.100}$$

The probability that a normally distributed variable X can be found within the interval (a,b) can be computed as

$$P\{X \in (a,b)\} = \int_a^b f(x)dx = \frac{1}{\sqrt{2\pi D_x}} \int_a^b \exp\left(\frac{(x-\bar{x})^2}{2D_x}\right)dx = \frac{1}{2}\left[\Phi\left(\frac{b-\bar{x}}{\sqrt{2D_x}}\right) - \Phi\left(\frac{a-\bar{x}}{\sqrt{2D_x}}\right)\right]. \tag{2.101}$$

Normal distribution is not only the most widespread one that is used in many practical problems, but it is also the *cornerstone of the entire probability theory.* This distribution arises when a random variable results from summation of a large number of independent or weakly dependent random variables comparable from the standpoint of their effect on the scattering of the sum. In other words, the distribution of the sum of a number of random variables converges, under very general conditions, to the normal distribution as the number of variables in the sum becomes large. This remarkable property of the normal distribution is known in the probability theory as the *central limit theorem.* This theorem states that the sum of many small random effects is normally distributed. The normal law has the maximum entropy (uncertainty) among all the laws of probability distributions of continuous random variables, for which only the expected (mean) value and the variance are known.

> **Example 2.24.** The under-keel clearance M for an oceangoing ship passing a shallow waterway when approaching a harbor in still water conditions is a normally distributed continuous random variable, with the following probability density function:
>
> $$f(m) = \frac{1}{\sqrt{2\pi D_m}} \exp\left[-\frac{(m-\bar{m})^2}{2D_m}\right], \tag{2.102}$$
>
> where $\bar{m}$ and D_m are the mean and the variance of the random variable M. The uncertainties associated with the under-keel clearance M are due to the fact that the ship's draft is affected by the errors in its prediction, possible ship's squat, uncertain water density, uncertainties in the charted water depth, sea bed relief, tide, siltation, and so on. Determine the safety factor $\gamma = \dfrac{\bar{m}}{\sqrt{2D_m}}$ that would enable the ship navigator to safely pass the waterway, if the required probability of safe passing is $P_s = 0.99999$.

The probability distribution function of the random variable M is

$$F(m) = \int_{-\infty}^{m} f(m)dm = \frac{1}{2}\left[1 + \Phi\left(\frac{m - \bar{m}}{\sqrt{2D_m}}\right)\right], \qquad (2.103)$$

where $\Phi(\alpha)$ is the Laplace function. The probability of safe passing can be computed as

$$P_s = 1 - F(0) = \left[1 + \Phi(\gamma)\right]. \qquad (2.104)$$

The calculated values of the safety factor γ for different probabilities P_s are shown in Table 2.3. As evident from the calculated data, the safety factor of $\gamma = 3.200$ corresponds to the required probability of safe passing the waterway. The approach used to solve this problem is typical and is widely used in engineering and applied science.

2.4.7 Rayleigh Distribution

If X and Y are normally distributed random variables, then the variable $R = \sqrt{X^2 + Y^2}$ is distributed in accordance with the Rayleigh law,

$$f(r) = \frac{r}{D_x}\exp\left[-\frac{r^2}{2D_x}\right], \qquad (2.105)$$

where the variance D_x is related to the variance D_r of the random variable R as $D_x = \frac{2}{4 + \pi}D_r$.

From (2.105), we find

$$F(r) = \int_0^r f(r)dr = 1 - \exp\left(-\frac{r^2}{2D_x}\right). \qquad (2.106)$$

The mean value, the variance, and the coefficient of variation of the Rayleigh distribution are

$$\bar{r} = \int_0^\infty rf(r)dr = \sqrt{\frac{\pi}{2}D_x}, \quad D_r = \int_0^\infty (r - \bar{r})^2 f(r)dr = \frac{4 + \pi}{2}D_x, \quad v_r = \frac{\sqrt{D_r}}{\bar{r}} = \sqrt{1 + \frac{4}{\pi}}. \qquad (2.107)$$

TABLE 2.3

Safety Factor in a Safe Passing a Shallow Waterway

P_s	0.9990000	0.9999000	0.9999900	0.9999990	0.9999999	1.0
γ	2.185	2.630	3.200	3.360	3.710	∞

Fundamentals of Applied Probability 43

Example 2.25. The random time T of human decision-making is distributed in accordance with the Rayleigh law:

$$f(t) = \frac{t}{D_x} \exp\left[-\frac{t^2}{2D_x}\right]. \tag{2.108}$$

How significant should the safety factor P_* be, so that the probability that the time of $t_* = 30$ s (required for making a decision in an extraordinary avionics safety-related situation) is exceeded with a very low probability $Q = 0.0001$?

The probability that a certain time t_* is exceeded, can be found as

$$Q = P(T \succ t_*) = \int_{t_*}^{\infty} f(t)dt = \int_{t_*}^{\infty} \frac{t}{D_x} \exp\left[-\frac{t^2}{2D_x}\right]dt = \exp\left(=\frac{t_*^2}{2D_x}\right). \tag{2.109}$$

This yields

$$D_x = -\frac{t_*^2}{2\ln Q} = 48.8581 \text{ s}^2; \quad D_x = \frac{4+\pi}{2} D_x = 174.4624 \text{ s}^2;$$

$$\bar{r} = \sqrt{\frac{\pi}{2}} D_x = 61.2345 \text{ s}^2.$$

The sought safety factor is

$$SF = -\frac{\bar{r}}{\sqrt{2D_r}} = \frac{61.2345}{\sqrt{2 \times 174.4624}} = 3.2782.$$

Example 2.26. The coefficient F of friction in an electronic module is a random variable that follows the Rayleigh distribution. Find the most likely value f_m (mode) of this coefficient and the probability that the actual F value will not exceed the limit f_m.

The maximum value of the Rayleigh curve (2.105) is $f_m = \sqrt{D_x}$. The sought probability is, therefore,

$$P\{F \prec f_m\} = \int_0^{f_m} \frac{f}{D_x} \exp\left(-\frac{f^2}{2D_x}\right)df = 1 - \exp\left(-\frac{f_m^2}{2D_x}\right) = 1 - \frac{1}{\sqrt{e}} = 0.3935.$$

Example 2.27. The random coefficient F of friction in a frictional-contact axial clutch is distributed in accordance with the Rayleigh law,

$$f_f(f) = \frac{f}{f_m^2} \exp\left(-\frac{f^2}{2f_m^2}\right), \tag{2.110}$$

where f_m is the most likely value (mode) of this coefficient. The measured axial force F_a in the clutch is related to the torque Q as $F_a = C\dfrac{\sqrt{Q}}{f}$, where the factor C (measured, as it follows from the last formula for the axial force, in $\sqrt{\dfrac{kg}{m}}$) depends on whether the uniform wear or a uniform pressure condition takes place. Establish the magnitude of the actuating force F_a in such a way that the probability that the actual torque Q will remain within the required limits $Q_1 \prec Q \prec Q_2$ is the largest. The torque Q is a nonrandom function of the random variable F, and its probability density function is expressed by the Rayleigh law

$$q_Q(Q) = \frac{Q}{Q_M^2}\exp\left(-\frac{Q^2}{2Q_m^2}\right),\tag{2.111}$$

where $Q_m = \left(\dfrac{f_m}{C}F_a\right)^2$ is the most probable value of the torque Q. The probability that the torque will be found within the required limits $Q_1 \prec Q \prec Q_2$ is

$$P\{Q_1 \prec Q \prec Q_2\} = \int_{Q_1}^{Q_2} \frac{Q}{Q_m^2}\exp\left(-\frac{Q^2}{2Q_m^2}\right)dQ = \exp\left(-\frac{Q_1^2}{2Q_m^2}\right) - \exp\left(-\frac{Q_2^2}{2Q_m^2}\right).\tag{2.112}$$

From this expression, we find that the probability P will be the largest if

$$Q_m = \frac{Q_2^2 - Q_1^2}{4\mathrm{In}\left(\dfrac{Q_2}{Q_1}\right)}.\tag{2.113}$$

In this case, the actuating force F_a assumes the value

$$F_a = C\frac{\sqrt{Q_m}}{f_m} = \frac{C}{2f_m}\frac{Q_2^2 - Q_1^2}{\mathrm{In}\left(\dfrac{Q_2}{Q_1}\right)} = \frac{CQ_1}{2f_m}\sqrt{\frac{\eta^2 - 1}{\mathrm{In}\,\eta}},\tag{2.114}$$

where $\eta = \dfrac{Q_2}{Q_1}$ is the ratio of the extreme torque values. If the actuating force is established in accordance with this formula, then there is a reason to expect that the torque will stay within the design limits $Q_1 \prec Q \prec Q_2$ during the operation of the clutch.

2.4.8 Weibull Distribution

The Rayleigh law can be obtained also as a special case of the following *Weibull distribution*:

Fundamentals of Applied Probability

$$f(x) = \frac{a}{b}\left(\frac{x}{b}\right)^{a-1} \exp\left[-\left(\frac{x-x_0}{b}\right)^a\right], \quad x \geq 0, a \succ 0, b \succ 0, x_0 \succ 0. \quad (2.115)$$

Here a is the *shape parameter*, b is the *scale parameter*, and x_0 is the *shift parameter* (the minimum possible value of the random variable X). The mean value and the variance of the random variable distributed in accordance with the Weibull law are

$$\bar{x} = b\Gamma\left(1+\frac{1}{a}\right) + x_0, \quad D_x = b^2\left[\Gamma\left(1+\frac{2}{a}\right) - \Gamma^2\left(1+\frac{1}{a}\right)\right], \quad (2.116)$$

where

$$\Gamma(\alpha) = \int_0^\infty x^{\alpha-1} e^{-x}\, dx \quad (2.117)$$

is the gamma-function. The Weibull cumulative distribution function is

$$F(x) = 1 - \exp\left[-\left(\frac{x-x_0}{b}\right)^a\right]. \quad (2.118)$$

When $x_0 = 0$ and $a = 2$, the Weibull law becomes the Rayleigh law; with $a = 1$ it becomes the exponential law; and with $a = 3$ it results in a distribution that is close to the normal distribution. The Weibull law is widely used in various engineering problems and particularly in reliability analyses and in the theory of material fatigue. The Weibull distribution is applicable also to the long-term distribution of ocean wave heights, extreme winds, and so on.

2.4.9 Beta-Distribution

Beta-distribution is defined over the range (a, b) as

$$f(x) = C(x-a)^\alpha (b-x)^\beta, \quad (2.119)$$

where both the parameters α and β are greater than –1. If α and β are integers, the constant C can be found from the condition of normalization as

$$C = \frac{(\alpha+\beta+1)!}{\alpha!\beta!(b-a)\alpha+\beta+1}. \quad (2.120)$$

The mean value and the variance of the beta-distribution are

$$\bar{x} = a + \frac{\alpha+1}{\alpha+\beta+2}(b-a), \quad D_x = \frac{(b-a)^2(\alpha+1)(\beta+1)}{(\alpha+\beta+2)^2(\alpha+\beta+3)}. \quad (2.121)$$

The beta-distribution is used in statistics and in reliability engineering. If, for a certain random variable, the mean, the variance, and the minimum and maximum values are available, then it is the beta-distribution that results in the maximum entropy (uncertainty). The beta-distribution will be revisited in connection with updating reliability in the three-step concept, when the recently suggested reliability physics oriented Boltzmann–Arrhenius–Zhurkov model is sandwiched between two statistical models—Bayes formula (for technical diagnostics applications) and beta-distribution (for updating reliability, when unexpected failures have been observed).

2.4.10 Beta-Distribution as a Tool for Updating Reliability Information

When updating reliability information, the estimated probability, P, of nonfailure can be treated itself as a random variable, and its probabilistic characteristics can be determined from the appropriate probability distribution. Beta-distribution is widely used for this purpose. The "successes" and "failures" in beta-distribution could be any physically meaningful (indicative) criteria. It could be, for example, the number of samples (including redundancies, whose probabilities of nonfailure are never 100%) that survived the accelerated tests, or, in the case of active redundancies, those that exhibited failures during such tests or in actual operation. It could be also the allowable levels of temperature, or the power of the dynamic response of the device to the input vibration power spectrum, or the level of the drop in the light output of a laser, or the level of leakage current. Such a generalization enables one to use the predictions based on the application of the beta-distribution, which is a powerful and flexible means for updating reliability.

The formal justification for using beta-distribution to update reliability information is based on the Bayesian formula, on one hand, and on the notion of conjugate distributions in this theory, on the other. If the posterior distribution of the probability of nonfailure remains in the same family as its prior distribution, then the prior and the posterior distributions are called *conjugate distributions*. For certain choices of the prior (conjugate prior), the posterior distribution has the same algebraic form as the prior, although generally with different parameter values. There are many distributions that are conjugate ones. When it comes to updating reliability, the two-parametric beta-distribution defined on the interval [0, 1] is considered the most easy to use. The beta-distribution is widely used in statistics. Two positive shape parameters, α and β, appear as exponents of the beta-distributed random variable X and control the shape of the distribution. The probability density function (2.119) of the beta-distribution can be also written as follows:

$$f(x) = \frac{x^{\alpha-1}(1-x)^{\beta-1}}{B(\alpha,\beta)}, \tag{2.122}$$

Fundamentals of Applied Probability 47

where

$$B(x,y) = B(y,x) = \int_0^1 t^{x-1}(1-t)^{y-1}\,dt = \frac{\Gamma(x)\Gamma(y)}{\Gamma(x+y)} = \frac{1}{y}\sum_0^\infty (-1)^n \frac{y(y-1)...(y-n)}{x!(x+n)}$$

(2.123)

is the beta-function, or Euler's integral of the first kind, and

$$\Gamma(x) = \int_0^\infty e^{-t} t^{x-1}\,dt$$

(2.124)

is the gamma-function (see [2.117]), or Euler's integral of the second kind. When n is an integer,

$$\Gamma(n) = (n-1)!$$

(2.125)

The probability density function (2.122) results in the cumulative distribution function

$$F(x;\alpha,\beta) = I_x(\alpha,\beta) = \frac{B(x;\alpha,\beta)}{B(\alpha,\beta)},$$

(2.126)

where

$$B(x;p,q) = B_x(p,q) = \int_0^x t^{p-1}(1-t)^{q-1}\,dt$$

(2.127)

is the incomplete beta-function and

$$I_x(p,q) = \frac{B_x(p,q)}{B(p,q)}$$

(2.128)

is the regularized incomplete beta-function.

The mean, the median, the mode, the variance, the skewness, and the kurtosis (which is the measure of the "peakedness" of the distribution) of the beta-distribution are as follows:

$$\prec x \succ = \frac{\alpha}{\alpha+\beta}, \quad Me \approx \frac{\alpha - \frac{1}{3}}{\alpha+\beta-\frac{2}{3}}, \text{ for } \alpha \succ 1, \beta \succ 1, \quad Mo = \frac{\alpha-1}{\alpha+\beta-2}, \text{ for } \alpha \succ 1, \beta \succ 1,$$

$$s^2 = D_x = \frac{\alpha\beta}{(\alpha+\beta)^2(\alpha+\beta+1)}, \quad \gamma_1 = \frac{2(\beta-\alpha)\sqrt{\alpha+\beta+1}}{(\alpha+\beta+2)\sqrt{\alpha\beta}},$$

$$\gamma_2 = \frac{6(\beta-\alpha)^2(\alpha+\beta+1) - \alpha\beta(\alpha+\beta+2)}{\alpha\beta(\alpha+\beta+2)(\alpha+\beta+3)}. \tag{2.129}$$

When the random variable of interest is probability, formula (2.122) can be written in the form

$$f(p) = \frac{p^{\bar{\alpha}}(1-p)^{\bar{\beta}}}{B(\bar{\alpha}+1, \bar{\beta}+1)}, \tag{2.130}$$

where the new parameters $\bar{\alpha}$ and $\bar{\beta}$ are introduced as $\bar{\alpha} = \alpha - 1$ and $\bar{\beta} = \beta - 1$. The denominator in this formula can be expressed through the gamma-function,

$$B(\bar{\alpha}+1, \bar{\beta}+1) = \frac{\Gamma(\bar{\alpha}+1)\Gamma(\bar{\beta}+1)}{\Gamma(\bar{\alpha}+\bar{\beta}+2)} = \frac{\bar{\alpha}!\bar{\beta}!}{(\bar{\alpha}+\bar{\beta}+1)!}, \tag{2.131}$$

and, therefore, the distribution (2.130) can be written also as

$$f(p) = \frac{(\bar{\alpha}+\bar{\beta}+1)!}{\bar{\alpha}!\bar{\beta}!} p^{\bar{\alpha}}(1-p)^{\bar{\beta}}. \tag{2.132}$$

Example 2.28. Let the performance of five "suspicious" (malfunctioning) devices be monitored, and one of them failed. Let us determine the beta-distribution characteristics for four successes and one failure $(\bar{\alpha}=4, \bar{\beta}=1)$. We have $\alpha = \bar{\alpha}+1=5$, $\beta = \bar{\beta}+1=2$, and the characteristics of the beta-distribution are

$$\text{mean}: \prec p \succ = \frac{\alpha}{\alpha+\beta} = \frac{5}{7} = 0.7143,$$

$$\text{variance}: s_p^2 = D_p = \frac{\alpha\beta}{(\alpha+\beta)^2(\alpha+\beta+1)} = \frac{10}{49\times 8} = 0.02551,$$

$$\text{median}: Me \approx \frac{\alpha - \frac{1}{3}}{\alpha+\beta-\frac{2}{3}} = \frac{\frac{14}{3}}{\frac{19}{3}} = 0.7368, \quad \text{mode}: M_o = \frac{\alpha-1}{\alpha+\beta-2} = \frac{4}{5} = 0.8,$$

$$\text{skewness}: \gamma_1 = \frac{2(\beta-\alpha)\sqrt{\alpha+\beta+1}}{(\alpha+\beta+2)\sqrt{\alpha\beta}} = \frac{-6\sqrt{8}}{9\sqrt{10}} = -0.5963,$$

$$\text{kurtosis}: \gamma_2 = \frac{6(\beta-\alpha)^2(\alpha+\beta+1) - \alpha\beta(\alpha+\beta+2)}{\alpha\beta(\alpha+\beta+2)(\alpha+\beta+3)} = \frac{6\times 9\times 8 - 90}{10\times 9\times 10} = 0.3800.$$

Fundamentals of Applied Probability

With $\alpha \succ \beta$ (there are more successes than failures), the distribution skews to the higher probabilities of nonfailure, and the mode (the maximum value, of the probability density function) is higher than the mean value and the median.

Let no failures be observed after the first two three-step concept steps have been carried out. Let us determine the expected number of successes (nonfailures) as a function of the probability of nonfailure. Assuming zero failures $\left(\bar{\beta}=0,\ \beta=1\right)$, we have $\bar{\alpha}=\dfrac{2\prec p \succ -1}{1-\prec p \succ}$. If the mean value of the probability of nonfailure is $\prec p \succ = 0.7143$, then $\bar{\alpha}=1.5002$. Since $\bar{\alpha}$ value has to be expressed by an integer, one should assume either $\bar{\alpha}=1\ (\alpha=2)$, or $\bar{\alpha}=2\ (\alpha=3)$. Then we obtain

$$\prec p \succ = \frac{\alpha}{\alpha+\beta} = \frac{2}{3} = 0.6667, \quad Me \approx \frac{\alpha-\dfrac{1}{3}}{\alpha+\beta-\dfrac{2}{3}} = \frac{\dfrac{5}{3}}{\dfrac{7}{3}} = 0.7143,$$

$$Mo = \frac{\alpha-1}{\alpha+\beta-2} = \frac{1}{1} = 1,$$

$$s_p^2 = \frac{\alpha\beta}{(\alpha+\beta)^2(\alpha+\beta+1)} = \frac{2}{9\times4} = 0.05556, \quad \gamma_1 = \frac{2(\beta-\alpha)\sqrt{\alpha+\beta+1}}{(\alpha+\beta+2)\sqrt{\alpha\beta}}$$

$$= \frac{-2\times2}{5\sqrt{2}} = -0.5657,$$

$$\gamma_2 = \frac{6(\beta-\alpha)^2(\alpha+\beta+1)-\alpha\beta(\alpha+\beta+2)}{\alpha\beta(\alpha+\beta+2)(\alpha+\beta+3)} = \frac{24-10}{2\times5\times6} = 0.2333,$$

in the case of $\alpha=2,\quad \beta=1$, and

$$\prec p \succ = \frac{\alpha}{\alpha+\beta} = \frac{3}{4} = 0.7500, \quad Me \approx \frac{\alpha-\dfrac{1}{3}}{\alpha+\beta-\dfrac{2}{3}} = \frac{\dfrac{8}{3}}{\dfrac{10}{3}} = 0.8000,$$

$$Mo = \frac{\alpha-1}{\alpha+\beta-2} = \frac{2}{2} = 1,$$

$$s_p^2 = D_p = \frac{\alpha\beta}{(\alpha+\beta)^2(\alpha+\beta+1)} = \frac{3}{16\times5} = 0.03750,$$

$$\gamma_1 = \frac{2(\beta-\alpha)\sqrt{\alpha+\beta+1}}{(\alpha+\beta+2)\sqrt{\alpha\beta}} = \frac{-4x\sqrt{5}}{6\sqrt{3}} = -0.8607,$$

$$\gamma_2 = \frac{6(\beta-\alpha)^2(\alpha+\beta+1)-\alpha\beta(\alpha+\beta+2)}{\alpha\beta(\alpha+\beta+2)(\alpha+\beta+3)} = \frac{24x5-18}{3x6x7} = 0.8095,$$

when $\alpha = 3$, $\beta = 1$. In both cases, it is a triangular distribution: the mode remains the same and is at $p = 1$. The mean and the median increase in the case $\alpha = 3$, $\beta = 1$ in comparison with the case $\alpha = 2$, $\beta = 1$, because of a larger number of successes. The variance reduces, because of the improved information, and the skewness (shift in the direction of higher probabilities of nonfailure) and the kurtosis ("peakedness") of the distribution increase.

Assume now that the predicted (anticipated) probability of nonfailure is as high as $\prec p \succ = 0.95$ Despite such a high probability of nonfailure, the product exhibited, nonetheless, a field failure. Let us determine, based on this additional information, the revised (updated) estimate of the actual operational probability of nonfailure. Assuming that the anticipated (projected) number of failures was zero $\left(\bar{\beta} = 0, \beta = 1\right)$ prior to putting the device or devices into operation, and using the formula for the number $\bar{\alpha}$ of anticipated nonfailures from the previous example, we obtain, with $\prec p \succ = 0.95$, that

$$\bar{\alpha} = \frac{2 \prec p \succ - 1}{1 - \prec p \succ} = \frac{0.90}{0.05} = 18.$$

For a new posterior failure, with $\bar{\alpha} = 18$ ($\alpha = 19$) and $\bar{\beta} = 1$ ($\beta = 2$), the revised characteristics of the beta-distribution for the probability of non-failure are

$$\prec p \succ = \frac{\alpha}{\alpha + \beta} = \frac{19}{21} = 0.9045, \quad Me \approx \frac{\alpha - \dfrac{1}{3}}{\alpha + \beta - \dfrac{2}{3}} = \frac{\dfrac{56}{3}}{\dfrac{61}{3}} = 0.9180,$$

$$Mo = \frac{\alpha - 1}{\alpha + \beta - 2} = \frac{18}{19} = 0.9474,$$

$$s_p^2 = D_p = \frac{\alpha\beta}{(\alpha + \beta)^2 (\alpha + \beta + 1)} = \frac{38}{441 \times 22} = 0.003917,$$

$$\gamma_1 = \frac{2(\beta - \alpha)\sqrt{\alpha + \beta + 1}}{(\alpha + \beta + 2)\sqrt{\alpha\beta}} = \frac{-34x\sqrt{22}}{23\sqrt{38}} = -1.1248,$$

$$\gamma_2 = \frac{6(\beta - \alpha)^2 (\alpha + \beta + 1) - \alpha\beta(\alpha + \beta + 2)}{\alpha\beta(\alpha + \beta + 2)(\alpha + \beta + 3)} = \frac{38148 - 874}{38 \times 23 \times 24} = 1.7770.$$

Thus, because of the occurrence of the unexpected failure, the actual probability of nonfailure of the product is only 90.45%, and not 95%. Note that this result, obtained assuming a 95% nonfailure level, indicates that after the first failure has occurred, as many as 19 additional continuous nonfailures (successes)—that is, $18 + 19 = 37$ successes and 1 failure—would have to be recorded (observed) in order to return the device's dependability (probability of nonfailure) to its original specified estimate of 95%.

Fundamentals of Applied Probability

We addressed previously a situation where one failure has occurred. Let us examine a situation with two failures. In this case, one should put $\bar{\alpha} = 18$ ($\alpha = 19$) and $\bar{\beta} = 2$ ($\beta = 3$), and the characteristics of the beta-distribution for the probability of nonfailure become as follows:

$$\prec p \succ = \frac{\alpha}{\alpha + \beta} = \frac{19}{22} = 0.8636, \quad Me \approx \frac{\alpha - \frac{1}{3}}{\alpha + \beta - \frac{2}{3}} = \frac{\frac{56}{3}}{\frac{64}{3}} = 0.8750,$$

$$Mo = \frac{\alpha - 1}{\alpha + \beta - 2} = \frac{18}{20} = 0.9000,$$

$$s_p^2 = D_p = \frac{\alpha \beta}{(\alpha + \beta)^2 (\alpha + \beta + 1)} = \frac{57}{484 \times 23} = 0.005120,$$

$$\gamma_1 = \frac{2(\beta - \alpha)\sqrt{\alpha + \beta + 1}}{(\alpha + \beta + 2)\sqrt{\alpha \beta}} = \frac{-32x\sqrt{23}}{24\sqrt{57}} = -0.8470,$$

$$\gamma_2 = \frac{6(\beta - \alpha)^2 (\alpha + \beta + 1) - \alpha \beta (\alpha + \beta + 2)}{\alpha \beta (\alpha + \beta + 2)(\alpha + \beta + 3)} = \frac{35328 - 1368}{57 \times 24 \times 25} = 0.9930.$$

Thus, the operational probability of nonfailure reduced by about 9.12%, compared to the projected probability of 95%, and by an additional 4.52% with respect to the situation with a single failure. The mean, the median, and the mode have also reduced, and because of the higher number of failures, the variance has increased, and the skewness and the kurtosis have decreased.

2.5 Functions of Random Variables

If X is a continuous random variable with a probability density function $f(x)$, and another continuous random variable Y is a function of the variable X, so that

$$Y = \varphi(X), \tag{2.133}$$

then the probability density function $g(y)$ of the random variable Y is expressed as

$$g(y) = f[\psi(y)]|\psi'(y)|, \tag{2.134}$$

where the function ψ is the inverse of φ, and the symbol $|\psi'(y)|$ indicates that the absolute value of the derivative $\psi'(y)$ of the function $\psi(y)$ should be taken. The rationale behind the formula (2.134) is that the "elementary" probability density functions $f(x)$ and $g(y)$ should be equal: $f_x(x)dx = g_y(y)dy$.

Example 2.29. The function Y is related to the argument X as $Y = \varphi(X) = e^{-\lambda X}$. If X is a normally distributed random variable,

$$f(x) = \frac{1}{\sqrt{2\pi D_x}} \exp\left(-\frac{(x-\bar{x})^2}{2D_x}\right),$$

what is the distribution of the random variable Y?

From $y = \varphi(x) = e^{-\lambda x}$ we find $x = \psi(y) = -\dfrac{\ln y}{\lambda}$, so that $\psi'(y) = -\dfrac{1}{\lambda y}$, and $|\psi'(y)| = \dfrac{1}{\lambda y}$. Then, with

$$f[\psi(y)] = \frac{1}{\sqrt{2\pi D_x}} \exp\left(-\frac{\left(-\dfrac{\ln y}{\lambda} - \bar{x}\right)^2}{2D_x}\right) = \frac{1}{\sqrt{2\pi D_x}} \exp\left(-\frac{\left(\dfrac{\ln y}{\lambda} + \bar{x}\right)^2}{2D_x}\right),$$

$$(2.135)$$

we have

$$g(y) = \frac{1}{\sqrt{2\pi D_x}} \frac{1}{\lambda y} \exp\left(-\frac{\left(\dfrac{\ln y}{\lambda} + \bar{x}\right)^2}{2D_x}\right). \qquad (2.136)$$

The probability density of the sum $Z = X + Y$ of two random variables is expressed as

$$g_z(z) = \int_{-\infty}^{\infty} f(x, z-x)dx = \int_{-\infty}^{\infty} f(z-y, y)dy, \qquad (2.137)$$

where $f(x, y)$ is the joint probability density function of the random variables X and Y. If these variables are statistically independent, then

$$f(x, y) = f_1(x)f_2(y), \qquad (2.138)$$

and the formula (2.126) can be written as

$$g_z(z) = \int_{-\infty}^{\infty} f_1(x)f_2(z-x)dx = \int_{-\infty}^{\infty} f_1(z-y)f_2(y)dy. \qquad (2.139)$$

The distribution of the sum $g(z)$ is called *composition or convolution of the distributions* $f_1(x)$ and $f_2(y)$. If the variables X and Y are statistically

Fundamentals of Applied Probability 53

independent, then the cumulative distribution function of the variable $Z = X + Y$ can be calculated as

$$G_z(z) = \int_{-\infty}^{z} f_1(x)dx = \int_{-\infty}^{z-x} f_2(y)dy = \int_{-\infty}^{z} f_1(x)F_2(z-x)dx, \qquad (2.140)$$

where

$$F_2(y)\int_{-\infty}^{y} f_2(y)dy \qquad (2.141)$$

is the cumulative distribution function of the variable Y.

The probability density of the random variable $Z = X - Y$ of two random variables is expressed as

$$g_z(z) = \int_{-\infty}^{\infty} f(x,x-z)dx = \int_{-\infty}^{\infty} f(z+y,y)dy, \qquad (2.142)$$

where $f(x, y)$ is the joint probability density function of the random variables X and Y. If these variables are statistically independent, then

$$g_z(z) = \int_{-\infty}^{\infty} f_1(x)f_2(x-z)dx = \int_{-\infty}^{\infty} f_1(z+y) f_2(y)dy. \qquad (2.143)$$

The distribution of the product $Z = XY$ is

$$g_z(z) = \int_{-\infty}^{\infty} \frac{f(x,z/x)}{|x|}dx = \int_{-\infty}^{\infty} \frac{f(z/y,y)}{|y|}dy. \qquad (2.144)$$

For statistically independent variables X and Y,

$$g_z(z) = \int_{-\infty}^{\infty} \frac{f_1(x)f_2(z/x)}{|x|}dx = \int_{-\infty}^{\infty} \frac{f_1(z/y)f_2(y)}{|y|}dy. \qquad (2.145)$$

Finally, the distribution of the random variable $Z = \dfrac{X}{Y}$ is

$$g_z(z) = \int_{-\infty}^{\infty} |y| f(zy,y)dy. \qquad (2.146)$$

For statistically independent variables X and Y,

$$g_z(z) = \int_{-\infty}^{\infty} |y| f_1(zy) f_2(y)dy. \qquad (2.147)$$

Example 2.30. A continuous random variable X is distributed in the interval $(0, 1)$ and has a probability density function $f(x) = 2x$. Find the mean value and variance of the function $Y = X^2$.

The mean and variance of the random variable X are

$$\bar{y} = m_2(x) = \int_0^1 x^2(2x)dx = \frac{1}{2};$$

$$D_y = m_2(y) - \bar{y}^2 = \int_0^1 (x^2)^2 (2x)dx - \left(\frac{1}{2}\right)^2 = \frac{1}{3} - \frac{1}{4} = \frac{1}{12}.$$

Example 2.31. A positive random variable X is characterized by the probability density function $f(x) = \lambda e^{-\lambda x}$ for $x \succ 0$. Find the mean, the variance, and the coefficient of variation (cov) of the random variable $Y = e^{-X}$.

The sought characteristics are

$$\bar{y} = \int_0^\infty e^{-x} \left(\lambda e^{-\lambda x}\right) dx = \frac{\lambda}{\lambda + 1},$$

$$D_y = m_2(y) - \bar{y}^2 = \int_0^\infty e^{-2x} \left(\lambda e^{-\lambda x}\right) dx - \left(\frac{\lambda}{\lambda + 1}\right)^2 = \frac{\lambda}{(\lambda + 1)(\lambda + 2)},$$

$$\mathrm{cov}_x = \frac{\sqrt{D_y}}{\bar{y}} = \sqrt{\frac{\lambda + 1}{\lambda(\lambda + 2)}}. \tag{2.148}$$

Example 2.32. Form a convolution of two exponential distributions $f_1(x_1) = \lambda_1 e^{-\lambda_1 x_1}$ and $f_2(x_2) = \lambda_2 e^{-\lambda_2 x_2}$, $x_1 \succ 0$, $x_2 \succ 0$.

The formula (2.123) yields

$$g(x) = \int_{-\infty}^\infty f_1(x_1) f_2(x - x_1)dx_1 = \int_0^x \lambda_1 e^{-\lambda_1 x_1} \lambda_2 e^{-\lambda_2(x - x_1)} dx_1$$

$$= \frac{\lambda_1 \lambda_2}{\lambda_2 - \lambda_1} \left(e^{-\lambda_1 x} - e^{-\lambda_2 x}\right), x \succ 0. \tag{2.149}$$

This distribution is known as Erlang's generalized distribution of order 2.

2.6 Extreme Value Distributions

The *extreme value* is defined as the largest value expected to occur in a certain limited number of observations (sample size) or during a certain (finite)

Fundamentals of Applied Probability 55

period of time. The number of observations or the time period should be established beforehand, prior to applying an extreme value evaluation technique. One should consider also, before this technique is applied, the response of the system or a human to the random loading at every cycle of loading (stressing) and determine the "regular," "ordinary," "nonextreme" probability density function $f(x)$ and the corresponding cumulative probability distribution function $F(x)$. These are "initial," "basic," "generating," "parent" functions. They characterize the response to the random excitation $X(t)$. The extreme response Y_n to the nth random loading is a random variable, and its probability density distribution $g(y_n)$ and cumulative probability distribution $G(y_n)$ are related to the basic distributions $f(x)$ and $F(x)$ by the following relationships:

$$g(y_n) = n\left\{ f(x)[F(x)]^{n-1} \right\}_{x=y_n} \tag{2.150}$$

and

$$G(y_n) = \left\{ [F(x)]^n \right\}_{x=\varsigma_n}, \tag{2.151}$$

respectively. These equations enable us to obtain, for sufficiently large n numbers, various statistical characteristics of the extreme values Y_n of the random process $X(t)$: the most probable extreme value (i.e., the value that is most likely to occur in n observations), the probability of exceeding a certain level, and so on.

In a simple reasoning, the formula (2.151) can be obtained using the rule of multiplying probabilities. Indeed, let n loadings be applied during a certain period of time t. Then the distribution of the absolute maxima of the process $X(t)$ of loading could be found as the maximum value of $X(t)$ in n series of observations with n observations in each series. Using the rule of multiplication of probabilities, we conclude that the cumulative probability density function for the absolute maximum in n loadings could be found as $G(x) = [F(x)]^n$ (i.e., by the formula [2.151]).

> **Example 2.33.** Let an aerospace device be operated in temperature cycling conditions, and the random amplitude of the induced stress, when a single cycle is applied, is distributed in accordance with the Rayleigh law:
>
> $$f(r) = \frac{r}{D_x} \exp\left(-\frac{r^2}{2D_x} \right).$$
>
> What is the most likely extreme value of the stress amplitude for a large number n of cycles?

The probability distribution function of the Rayleigh law is

$$F(r) = \int_0^r f(r)\,dr = 1 - \exp\left(-\frac{r^2}{2D_x}\right).$$

Then the probability distribution density function of the extreme random value Y_n of the stress amplitude can be found using the formula (2.150) as

$$g\left(y_n\right) = n\varsigma_n^2 \exp\left(-\frac{\varsigma_n^2}{2}\right)\left[1 - \exp\left(-\frac{\varsigma_n^2}{2}\right)\right]^{n-1}, \qquad (2.152)$$

where $\varsigma_n = \dfrac{y_n}{\sqrt{D_x}}$ is the sought dimensionless amplitude. Its maximum value could be determined from the equation $g'\left(y_n\right) = 0$, which yields

$$\varsigma_n^2\left[n\exp\left(-\frac{\varsigma_n^2}{2}\right) - 1\right] - \left[\exp\left(-\frac{\varsigma_n^2}{2}\right) - 1\right] = 0. \qquad (2.153)$$

If the number n is large, the second term in this expression is small and can be omitted, so that

$$n\exp\left(-\frac{\varsigma_n^2}{2}\right) - 1 = 0, \qquad (2.154)$$

and

$$y_n = \varsigma_n\sqrt{D_x} = \sqrt{2D_x \ln n}. \qquad (2.155)$$

As evident from this result, the ratio $\dfrac{y_n}{\sqrt{D_x}}$ of the extreme response y_n of the device experiencing temperature cycling after n cycles are applied to the standard deviation of the response when a single cycle is applied, is $\sqrt{2\ln n}$. This ratio is 3.2552 for 200 cycles, 3.7169 for 1000 cycles, and 4.1273 for 5000 cycles. One encounters a similar situation, when a human has to make many times a go/non-go decision, and the decision-making time is a random variable distributed in accordance with the Rayleigh law (see Chapter 3).

3

Helicopter-Landing-Ship and the Role of the Human Factor

3.1 Summary

We address, using probabilistic modeling and the extreme value distribution (EVD) technique, the helicopter undercarriage strength in a helicopter-landing-ship (HLS) situation. Since nothing and nobody is 100% predictable, our analysis contains an attempt to quantify, on the probabilistic basis, the role of a human factor in the situation in question. This factor is important from the standpoint of the operation time that affects the likelihood of safe landing (actually helicopter undercarriage strength) during the relatively short lull period in the sea condition. The operation time includes the time required for the officer on board and the helicopter pilot to make their go-ahead decisions, and the time of actual landing. It is assumed, for the sake of simplicity, that both of these random times could be approximated by the Rayleigh's law, while the random lull duration follows the normal law. This law is characterized here by a high ratio of the mean value to the standard deviation. Then negative values of the normally distributed time of the lull are next to zero and do not play a role.

Safe landing could be expected if the probability that it occurs during the lull time is sufficiently high. The probability that the helicopter's undercarriage strength is not compromised can be evaluated as a product of the probability that landing occurs during the lull time and the probability that the relative velocity of the helicopter undercarriage with respect to the ship's deck at the moment of landing does not exceed the allowable level. This level is supposed to be determined beforehand based on the available allowable landing velocity in the helicopter-landing-ground condition.

The developed probabilistic model can be used when developing specifications for the helicopter undercarriage strength, as well as guidelines for personnel training. Particularly, the model can be of help when establishing the time to be met by the two humans making their go-ahead decisions for landing and the time to actually land the helicopter. Plenty of additional risk analyses (associated with the need to quantify various underlying

57

3.2 Introduction

Human error contributes to about 80% of vehicular (maritime, avionic, and even automotive) casualties and accidents [1–4]. This large percentage should not be attributed, of course, to the direct human error only. A mishap often occurs because an erroneous decision is made by the vehicle operator in conditions of uncertainty as a result of his/her interactions, in varying environmental conditions, with never-perfect forecasts, never 100% dependable navigation instrumentation and operation equipment, uncertain and often harsh environments, and not-always-user-friendly information.

Considerable improvement in various safety-at-sea, safety-in-air, and safety-in-the-outer-space situations can be expected through better ergonomics, better work environment, better training, and other aspects directly affecting human behavior in hazardous situations (see, for instance, [5–7]), as well as through improving the sensitivity, accuracy, and robustness of the navigation devices and the operation equipment. There is also a significant opportunity (potential) for further reducing casualties and accidents at sea and in air through better understanding the role that various uncertainties play in the operator's world of work. Uncertainties affecting the safe operation of a vehicle are associated with the human factor both directly (human fatigue, delayed reaction, erroneous decision-making, etc.) and indirectly, because of the imperfect forecast or because of the human interaction with various imperfect instrumentation and equipment.

The major uncertainties include, but might not be limited to, the instrumentation and equipment performance, environmental conditions, accuracy and consistency of the processed information, predictability and timeliness of the response of the vehicle (object of control) to the harsh environmental conditions (rough seas, winds, currents, tides, gusts, atmospheric turbulence, outer space radiation, etc.) and/or to the operator's actions. Obviously, not all these uncertainties have to be considered in each safety-related situation, and not all the uncertainties have to be accounted for on the probabilistic basis, but it is also obvious that a careful insight into the possible and critical uncertainties has to be developed whenever appropriate and possible.

By employing measurable ways to assess the role, the contributions, and the interaction ("interfaces") of the various uncertain factors with the on-board hardware and software, one could improve significantly the safety-at-sea, in the air, and in the outer space, including human performance. This performance should be often viewed as an important (perhaps the most important) part of the complex "man-instrumentation-vehicle-environment" system.

Helicopter-Landing-Ship and the Role of the Human Factor

Such a "systemic" approach, if properly developed and implemented, would enable one to predict and, if necessary, effectively minimize, with the highest cost-effectiveness possible, the probability of occurrence of a casualty or an accident at sea, in the air, or in the outer space.

The analysis that follows contains an attempt to quantify, on the probabilistic basis [8], the role that the human factor plays, in terms of the reaction (decision-making) time of the officer on ship board and helicopter pilot, in a HLS situation. The developed model can be used, with proper modifications, to analyze the vertical take-off and landing (VTOL) situation [9–17] as well. NASA-Ames Center has conducted in the past evaluations of the navigation performance of shipboard-VTOL-landing guidance systems and several piloted simulation evaluations to assess the merits of a predictive lull swell guidance law for landing a VTOL aircraft at sea (see, for instance, [11–13]). Blackwell and Feik [14] developed and implemented on an ELXSI 6400 computer a mathematical model of the on-deck helicopter/ship dynamic interface. This work provides a capability for investigating helicopter/ship dynamic interactions, such as deck clearances on landing, swaying, toppling and sliding criteria, and tie-down loads. Different helicopter types can be readily examined and compared, given their undercarriage representation. Thomson, Coton, and Galbraith [15] carried out a simulator-based study of helicopter-ship-landing procedures incorporating measured flow data. Carico and Ferrier [16] designed a Landing Period Designator to provide ship motion cues to the pilot to assist him/her in anticipating ship deck quiescent periods that result in acceptable conditions for a shipboard landing. Suhir [8] applied the EVD technique to evaluate the helicopter undercarriage strength when landing on a solid ground and on a ship deck. He showed also [18] how this technique could be applied to some other safety-at-sea situations, such as, for example, establishing the adequate under-keel clearances for a large crude-oil carrier entering harbor. It is the HLS problem that is addressed in this chapter.

3.3 Probability That the Operation Time Exceeds a Certain Level

The highest safety, in terms of the helicopter undercarriage strength, when landing on a ship deck, could be expected, if such landing occurs during the lull period of the seas. Typically, the officer on ship board, using the information from the on-board surveillance systems, signals to the helicopter pilot, when the lull period ("wave window") commences. The challenge is, of course, to foresee, to an extent possible, the duration of the lull. If the (random) sum, $T = t + \theta$, of the (random) time, t, needed for the officer on ship board and the helicopter pilot to make their go-ahead decisions on landing, and the (random) time, θ, needed to actually land the helicopter on the ship's deck, is lower, with a high probability, than the (random) duration, L, of the lull, then safe landing becomes possible. In the analysis that follows, we assume the simplest probability distributions for the random time periods of interest. We use the single parametric Rayleigh distributions

$$f_t(t) = \frac{t}{t_0^2} \exp\left(-\frac{t^2}{2t_0^2}\right), \quad f_\theta(t) = \frac{\theta}{\theta_0^2} \exp\left(-\frac{\theta^2}{2\theta_0^2}\right), \tag{3.1}$$

as suitable approximations for the random times t and θ of decision-making (by both humans) and landing, respectively, and the normal law

$$f_l(l) = \frac{1}{\sqrt{2\pi}\sigma} \exp\left(-\frac{(l-l_0)^2}{2\sigma^2}\right), \quad \frac{l_0}{\sigma} \geq 4.0, \tag{3.2}$$

as a suitable approximation for the random duration, L, of the lull. In the formulas (3.1) and (3.2), t_0 and θ_0 are the most likely times of decision-making and landing (maxima of the functions [3.1]), respectively (in the case of a Rayleigh law these times coincide with the standard deviations of the random variables in question), l_0 is the most likely (mean) value of the normally distributed lull time, and σ is the standard deviation of this time. The ratio $\frac{l_0}{\sigma}$ ("safety factor") of the mean value of the lull time to its standard deviation should be large enough (say, larger than 4) to make the normal law usable for the application in question: the random time cannot be negative, and if the ratio $\frac{l_0}{\sigma}$ is significant, the negative ordinates of the normal distribution are negligible and can be ignored. The probability, P_*, that the sum $T = t + \theta$ of the random variables t and θ, which is the total time of the two decision-making times and the time of actual landing, exceeds an arbitrary time level, $\hat{T}$, can be found, using the rules of convolution of distributions (see, e.g., Ref. [8]), as follows:

$$P_* = 1 - \int_0^{\hat{T}} \frac{t}{t_0^2} \exp\left(-\frac{t^2}{2t_0^2}\right) \left[1 - \exp\left(-\frac{(T-t)^2}{2\theta_0^2}\right)\right] dt = \exp\left(-\frac{\hat{T}^2}{2t_0^2}\right)$$

$$+ \exp\left[-\frac{\hat{T}^2}{2(t_0^2 + \theta_0^2)}\right] \left\{\frac{\theta_0^2}{t_0^2 + \theta_0^2} \left[\exp\left[-\frac{t_0^2 \hat{T}^2}{2\theta_0^2(t_0^2 + \theta_0^2)}\right]\right] - \exp\left[-\frac{\theta_0^2 \hat{T}^2}{2t_0^2(t_0^2 + \theta_0^2)}\right]\right\}$$

$$+ \sqrt{\frac{\pi}{2}} \frac{\hat{T} t_0 \theta_0}{(t_0^2 + \theta_0^2)^{3/2}} \exp\left[-\frac{\hat{T}^2}{2(t_0^2 + \theta_0^2)}\right] \left\{\left[\Phi\left[\frac{t_0 \hat{T}}{\theta_0 \sqrt{2(t_0^2 + \theta_0^2)}}\right]\right] + \Phi\left[\frac{\theta_0 \hat{T}}{t_0 \sqrt{2(t_0^2 + \theta_0^2)}}\right]\right\}$$

$$\tag{3.3}$$

where

$$\Phi(x) = \frac{2}{\sqrt{\pi}} \int_0^x e^{-z^2}\, dz \qquad (3.4)$$

is the Laplace function (see [2.94]). Clearly, when the time $\hat{T}$ is zero, this time will always be exceeded ($P_* = 1$). When the time $\hat{T}$ is infinitely long $(\hat{T} \to \infty)$, the probability that this time is exceeded is always zero ($P_* = 0$). When the most likely duration of landing, θ_0, is very small compared to the most likely time, t_0, required for making the two go-ahead decisions, the expression (3.3) yields

$$P_* = \exp\left(-\frac{\hat{T}^2}{2t_0^2}\right) \qquad (3.5)$$

In this case, the probability that the total time of operation exceeds a certain time duration, $\hat{T}$, depends only on the most likely time, t_0, of decision-making. From (3.5), we have

$$\frac{t_0}{\hat{T}} = \frac{1}{\sqrt{-2\ln P_*}}. \qquad (3.6)$$

If the acceptable probability, P_*, of exceeding the time, $\hat{T}$ (e.g., the duration of the lull, if this duration is treated as a nonrandom variable of the level $\hat{T}$), is, say, $P = 10^{-4} = 0.01\%$, then the time of making the go-ahead decisions should not exceed $0.2330 = 23.3\%$ of the time, $\hat{T}$ (the expected duration of the lull), otherwise the requirement $P \le 10^{-4} = 0.01\%$ will be compromised. Similarly, when the most likely duration, t_0, of decision-making is very small compared to the most likely time, θ_0, of actual landing, the formula (3.3) yields

$$P_* = \exp\left(-\frac{\hat{T}^2}{2\theta_0^2}\right) \qquad (3.7)$$

(i.e., the probability of exceeding a certain time level $\hat{T}$ depends only on the most likely time θ_0 of landing).

As follows from the formulas (3.1), the probability that the actual time of decision-making or the time of landing exceeds the corresponding most likely time is expressed by the formulas of the types (3.5) and (3.7), and, as long as the Rayleigh distribution is used, is as high as $P_* = \frac{1}{\sqrt{e}} = 0.6065 = 60.6\%$. In this connection we would like to mention that the one-parametric Rayleigh law is characterized by a rather large standard deviation and therefore might not be the best approximation for the probability density functions for the decision-making time and the time of landing. A more "powerful" and a more flexible two-parametric law, such as, for example, the Weibull law, might be more suitable and more practical as an appropriate probability distribution of the random times, t and θ. Its use, however, will make our analysis unnecessarily more complicated, and our goal is not so much, as they say, to "dot all the i's and cross all the t's," as far as modeling of the role of the human factor in the problem in question is concerned, but rather to demonstrate that the attempt to use methods of the applied probability to quantify the role of the human factor in some safety-at-sea and similar problems in aerospace engineering might be quite fruitful. When developing practical guidelines and recommendations, a particular law of the probability distribution should be established based on the actual statistical data, and employment of various goodness-of-fit criteria might be needed in detailed statistical analyses. These are, however, beyond the scope of this study.

When the most likely times t_0 and θ_0 required for making the go-ahead decisions and for the actual landing, are equal, the formula (3.3) yields:

$$P_* = P_*\left(\frac{t_0}{\hat{T}}, \frac{\theta_0}{\hat{T}}\right) = \exp\left(-\frac{\hat{T}^2}{2t_0^2}\right)\left[1 + \sqrt{\pi}\,\frac{\hat{T}}{2t_0}\exp\left(\left(\frac{\hat{T}}{2t_0}\right)^2\right)\Phi\left(\frac{\hat{T}}{2t_0}\right)\right]. \quad (3.8)$$

For large enough $\dfrac{\hat{T}}{t_0}$ ratios $\left(\dfrac{\hat{T}}{t_0} \geq 3\right)$, the second term in the brackets becomes large compared to unity, so that only this term should be considered. The calculated probabilities of exceeding a certain time level, $\hat{T}$, based on the formula (3.8), are shown in Table 3.1. In the third row of this table, we indicate, for the sake of comparison, the probabilities, P°, of exceeding the given time, $\hat{T}$, when only the time t_0 or only the time θ_0 is different from zero (i.e., for the special case that is mostly remote from the case $t_0 = \theta_0$). Clearly, the

TABLE 3.1

The Probability P_* That the Operation Time Exceeds a Certain Time Level $\hat{T}$ versus the Ratio $\hat{T}/t_0$ of This Time Level to the Most Likely Time t_0 of Decision-Making for the Case When the Time t_0 and the Most Likely Time θ_0 of Actual Landing Are the Same

$\hat{T}/t_0$	6	5	4	3	2
P_*	6.562E-4	8.553E-3	6.495E-2	1.914E-1	6.837E-1
P°	1.523E-8	0.373E-5	0.335E-3	1.111E-2	1.353E-1
P_*/P°	4.309E4	2.293E3	1.939E2	1.723E1	5.053

For the sake of comparison, the probability P° of exceeding the time level $\hat{T}$, when either the time t_0 or the time θ_0 are zero, is also indicated.

Helicopter-Landing-Ship and the Role of the Human Factor

probabilities computed for other possible combinations of the times t_0 and θ_0 could be found between the calculated probabilities P_* and $P°$.

The following conclusions can be drawn from Table 3.1 data:

1. The probability that the total time of operation (the time of decision-making and the time of landing) exceeds the given time level $\hat{T}$ rapidly increases with an increase in the time of operation.

2. The probability of exceeding the time level $\hat{T}$ is considerably higher, when the most likely times of decision-making and of landing are finite, and particularly are equal to each other, in comparison with the situation when one of these times is significantly shorter than the other one (i.e., zero or next-to-zero). This is especially true for short operation times: the ratio $P_*/P°$ of the probability P_* of exceeding the time level $\hat{T}$ in the case of $t_0 = \theta_0$ to the probability $P°$ of exceeding this level in the case $t_0 = 0$ or in the case $\theta_0 = 0$ decreases rapidly with an increase in the total time of operation. Thus, there exists a significant incentive for reducing the operation time. The importance of this intuitively obvious fact is quantitatively assessed by the conducted analysis.

The data of the type shown in Table 3.1 can be used, particularly, to train the personnel for a quick reaction in the situation in question. If, for instance, the expected duration of the lull is 30 s, and the required (specified) probability of exceeding this time is $P = 10^{-3}$, then, as evident from the table data, the times for decision-making and actual landing should not exceed 5.04 s. It is advisable, of course, that these predictions are verified by simulation and by actual practices. The statistical information about the lull durations could be obtained from the short-term forecasts for the given area of the ocean [19]. Other useful information that could be drawn from the data of the type shown in Table 3.1 is whether it is possible at all to train a human to react (make a decision) in just a couple of seconds. If not, then one should decide on a broader employment of more sophisticated, more powerful, more flexible, and more expensive equipment to do the job. If pursuing such an effort is decided upon, then probabilistic sensitivity analyses of the type carried out in this chapter will be needed to determine the most promising ways to go.

3.4 Probability That the Duration of Landing Exceeds the Duration of the Lull

The lull time L is a random normally distributed variable, and the probability that this time is found below a certain level $\hat{L}$ is

$$P_l = P_l\left(\frac{\sigma}{\hat{L}}, \frac{l_0}{\hat{L}}\right) = \int_{-\infty}^{\hat{L}} f_l(l)\,dl = \frac{1}{2}\left[1 + \Phi\left(\frac{\hat{L} - l_0}{\sqrt{2}\sigma}\right)\right] = \left[1 + \Phi\left(\frac{1 - \dfrac{l_0}{\hat{L}}}{\sqrt{2}\dfrac{\sigma}{\hat{L}}}\right)\right]. \tag{3.9}$$

The probability that the lull time in the situation in question is exceeded can be determined by equating the times $\hat{T} = \hat{L} = T$ and computing the product

$$P_A = P_*\left(\frac{t_0}{T}, \frac{\theta_0}{T}\right)P_l\left(\frac{\sigma}{T}, \frac{l_0}{T}\right) \tag{3.10}$$

of the probability, $P_*\left(\dfrac{t_0}{T}, \dfrac{\theta_0}{T}\right)$, that the time of operation exceeds a certain level, T, and the probability, $P_l\left(\dfrac{\sigma}{T}, \dfrac{l_0}{T}\right)$, that the duration of the lull is shorter than this time. Formula (3.10) considers the effect of the sea conditions (through the values of the most likely duration, l_0, of the random lull time, L, and its standard deviation, σ), the role of the human factor, t_0 (the total most likely time required for the officer on ship board and the helicopter pilot to make their go-ahead decisions for landing), and the most likely time, θ_0, of actual landing (which characterizes both the qualification of the helicopter pilot and the qualities/behavior of the flying machine) on the probability of safe landing. After a low enough allowable value, P_A^*, of the probability, P_A, is established and agreed upon, Equation (3.10) can be used to determine the allowable maximum most likely time, θ_0, of landing. The actual time of landing can be assessed by the formula of the type (3.6):

$$\Delta t^* = \theta_0\sqrt{-2\ln P_l}, \tag{3.11}$$

where P_l is the allowable probability that the level Δt^* is exceeded. If, for instance, $\theta_0 = 10\,\mathrm{s}$ and $P_l = 0.00001$, then the time of landing is $\Delta t^* = 48.0\,\mathrm{s}$.

3.5 The Probability Distribution Function for the Extreme Vertical Velocity of Ship's Deck

The cumulative probability distribution function for the extreme vertical ship velocity $\dot{Z}^*$ (the probability that the vertical velocity of the ship deck at the HLS location is below a certain level $\dot{Z}^*$) due to her motions in waves can be expressed, using the EVD technique, as follows [3–8]:

$$F_{\dot{z}^*}\left(\dot{z}^*\right) = \exp\left[-n^*\exp\left(-\frac{\left(\dot{z}^*\right)^2}{2D_{\dot{z}}}\right)\right] - \exp\left(-n^*\right). \tag{3.12}$$

Here $D_{\dot{z}}$ is the variance of the ship's vertical random velocity $\dot{z}$; $n^* = \dfrac{\Delta t^*}{\tau_e}$ is the expected number of ship oscillations during the helicopter's landing time Δt^*; and τ_e is the effective period of the ship motion in irregular seas. The formula $n^* = \dfrac{\Delta t^*}{\tau_e}$ for the n^* value reflects an assumption that a ship in irregular waves behaves as a sort of a narrow-band filter that enhances the oscillations whose frequencies are close to the ship's own natural frequency (in still water) in her heave and pitch (as is known, these two frequencies are rather close) and suppresses all the other frequencies.

Examine special cases for the expression (3.12). If the level $\dot{z}^*$ is zero, the formula (3.12) yields $F_{\dot{z}^*}(0) = 0$. If the landing time (measured by the expected number n^* of ship oscillations) is significant, then the second term in (3.12) becomes small, and this formula can be simplified:

$$F_{\dot{z}^*}(\dot{z}^*) \approx \exp\left[-n^* \exp\left(-\frac{(\dot{z}^*)^2}{2D_{\dot{z}}}\right)\right].\tag{3.13}$$

If, in such a situation, the level $\dot{z}^*$ is zero, the function $F_{\dot{z}^*}(\dot{z}^*)$ becomes

$$F_{\dot{z}^*}(0) = \exp(-n^*)\tag{3.14}$$

and, for a high enough n^* value, one still obtains $F_{\dot{z}^*}(0) = 0$. If, however, for a finite n^* (which is never zero and cannot be smaller than one) the level $\dot{z}^*$ is high, the function $F_{\dot{z}^*}(\dot{z}^*)$ becomes $F_{\dot{z}^*}(\infty) = 1$, as it is supposed to be.

3.6 Allowable Landing Velocity When Landing on a Solid Ground

The landing velocity, V, when landing on a solid ground, is a random variable that could be assumed, based on physical and common sense considerations, to be normally distributed:

$$f_v(v) = \frac{1}{\sqrt{2\pi D_v}} \exp\left[-\frac{(v-\bar{v})^2}{2D_v}\right],\tag{3.15}$$

where $\bar{v}$ is the mean value of the velocity V, and D_v is its variance. The probability distribution function of this velocity (i.e., the probability that the random velocity V is below a certain value v) is

$$F_v(v) = \frac{1}{2}\left[1 + \Phi\left(\frac{v - \bar{v}}{\sqrt{2D_v}}\right)\right].$$ (3.16)

The allowable level v^* of the landing velocity V, assuming a large enough probability $F_v(v^*)$, can be found from the equation

$$F_v(v^*) = \frac{1}{2}\left[1 + \Phi\left(\frac{v^* - \bar{v}}{\sqrt{2D_v}}\right)\right].$$ (3.17)

3.7 Allowable Landing Velocity When Landing on a Ship Deck

The cumulative distribution function for the relative vertical velocity

$$V_r = V + \dot{Z}^*$$ (3.18)

of the helicopter with respect to the ship's deck can be evaluated as

$$F(v) = P(V_r \le v) = \int_{-\infty}^{v_r} f_v(v) F_{\dot{z}^*}(v_r - v)\,dv$$

$$= \frac{1}{\sqrt{2\pi D_v}} \int_{-\infty}^{v_r} \exp\left[-\frac{(v - \bar{v})^2}{2D_v}\right]\left[\exp\left(-n^* \exp\left(-\frac{(\dot{z}_*)^2}{2D_{\dot{z}}}\right)\right) - \exp(-n^*)\right]dv$$

$$= \frac{1}{\sqrt{\pi}} \int_{0}^{\infty} \exp\left[-(\xi - \gamma)^2\right]\left[\exp\left(-n^* \exp(-\delta\xi^2)\right) - \exp(-n^*)\right]d\xi,$$ (3.19)

where $\xi = \dfrac{v}{\sqrt{2D_v}}$ is the variable of integration (a sort of safety factor for the landing velocity);

$$\gamma = \frac{v_r - \bar{v}}{\sqrt{2D_v}} = \gamma_t - \gamma_v$$ (3.20)

is the safety factor associated with the ship motions, which is computed as the difference between the total safety factor

$$\gamma_t = \frac{v_r}{\sqrt{2D_v}},$$ (3.21)

Helicopter-Landing-Ship and the Role of the Human Factor 67

when landing in rough seas on the ship's deck, and the safety factor

$$\gamma_v = \frac{\bar{v}}{\sqrt{2D_v}},$$ (3.22)

when landing on the solid ground; and

$$\delta = \frac{D_r}{D_{\dot{z}}}$$ (3.23)

is the ratio of the variance, D_r, of the relative velocity, V_r, of the helicopter undercarriage with respect to the ship's deck to the variance, $D_{\dot{z}}$, of the ship's vertical velocity $\dot{z}$. The formula (3.18) determines the probability that the random relative velocity, V_r, of the helicopter undercarriage with respect to the ship's deck remains below a certain value, v_r. When $D_{\dot{z}} \to \infty$ (significant ship motions) and/or $D_r \to 0$ (insignificant absolute vertical velocities of the helicopter), the ratio $\delta = \dfrac{D_r}{D_{\dot{z}}} \to 0$. This situation is unfavorable for the under-carriage strength: the probability that the extreme vertical velocity of the helicopter during its landing on the ship's deck remains below a certain v value is zero: $F(v) = 0$. For large enough (but not very large) n^* values (landing lasts for a rather long time), the formula (3.19) yields

$$F(v) = \frac{1}{\sqrt{\pi}} \int_0^\infty \exp\left[-(\xi - \gamma)^2 - n^* \exp\left(-\delta \xi^2\right) \right] d\xi.$$ (3.24)

For very large n^* values, we have $F(v) = 0$. Such a situation is also unfavorable for safe landing. For not very large n^* values, however (landing does not take long), but large $\delta = \dfrac{D_r}{D_{\dot{z}}}$ ratios (significant variance of the relative velocity, but insignificant variance of the velocity of the vertical ship motions), the formula (3.24) can be further simplified:

$$F(v) = \frac{1}{\sqrt{\pi}} \int_0^\infty \exp\left[-(\xi - \gamma)^2 \right] d\varsigma = \frac{1}{2}(1 + \Phi(\gamma)) = \frac{1}{2}\left[1 + \Phi\left(\frac{v - \bar{v}}{\sqrt{2D_v}} \right) \right].$$ (3.25)

This formula is not (and should not be) different from the formula (3.17) for the case of safe landing on a solid ground. For small $\delta = \dfrac{D_r}{D_{\dot{z}}}$ ratios (but still large n^* values), the formula (3.24) yields

$$f(v) = \frac{1}{2}\exp(-n^*)\left[1 + \Phi\left(\frac{v - \bar{v}}{\sqrt{2D_v}} \right) \right]$$ (3.26)

This formula contains a factor $\exp(-n^*)$ that accounts for the finite duration of landing. When n^* is very small (very short time of landing), the situation is not different from the case of landing on a solid ground. When n^* is large, the situation is certainly unfavorable: $f(v) = 0$. Thus, the probability that a certain level v_* of the relative velocity V_r of the helicopter with respect to the ship's deck is not exceeded can be found as $P_B = F(v_*)$. The probability $F(v)$ can be determined using either the general formula (3.19), or one of the formulas (3.24), (3.25), or (3.26) for a particular special case.

3.8 The Probability of Safe Landing on a Ship's Deck

The probability P_C that the undercarriage strength will not be compromised during helicopter landing (probability of safe landing) can be evaluated as a product of the probability $1 - P_A$ that the helicopter will be able to land during the lull time and the probability $P_B = F(v_*)$ that the relative velocity of the helicopter with respect to the ship's deck will not exceed a certain allowable (specified) level v^*:

$$P_C = (1 - P_A)P_B. \tag{3.27}$$

If the landing velocity, v_0, on the ground is treated as a deterministic value (if the variance D_v of this velocity can be considered zero) and the allowable relative velocity v^* (which is due to the undercarriage structure only) are known, then the condition of safe landing becomes quite simple. Indeed, in such a situation, Equation (3.12) results in the following simple formula for the extreme value $\dot{z}^*$ of the ship's vertical velocity:

$$\dot{z}^* = \sqrt{2D_{\dot{z}}\left[\ln n^* - \ln\left(-\ln P_D + \exp(-n^*)\right)\right]}, \tag{3.28}$$

and the condition of safe landing becomes

$$\dot{z}^* \le v^* - v_0. \tag{3.29}$$

> **Example 3.1.** Let the most likely times of the go-ahead decision-making and of the actual landing be the same and equal to $t_0 = \theta_0 = 10\,\text{s}$, the most likely (mean) lull time be $l_0 = 20\,\text{s}$, and the standard deviation of the lull time be $\sigma = 5\,\text{s}$. Then, using the formulas (3.9) and (3.10), and the data in Table 3.1, we obtain the values in Table 3.2.

Helicopter-Landing-Ship and the Role of the Human Factor

TABLE 3.2

The Probability P_A of Safe Landing versus the Ratio T/t_0 of the Normally Distributed Duration T of the Lull to the Most Likely Time t_0 of Decision-Making or the Most Likely Time θ_0 of Actual Landing, When the Times t_0 and θ_0 Are Equal

T/t_0	6	5	4	3	2
P_i	6.562E-4	8.553E-3	6.495E-2	1.914E-1	6.837E-1
T/l_0	3.0	2.5	2.0	1.5	1.0
P_l	1.0	1.0	0.9999	0.9770	0.5000
P_A	6.562E-4	8.553E-3	6.494E-2	1.870E-1	3.418E-1

As evident from Table 3.2 data, the probability P_A that the time of operations exceeds the duration of the lull increases rapidly with the decrease in the ratio of the lull duration to the most likely time of either the decision-making or the landing process, while the probability that the lull duration is below a certain value decreases with the decrease in the ratio of this value to the most likely lull duration. The first effect prevails, and the product of these two probabilities (defining the likelihood that the helicopter is not successful in landing on the ship's deck during the lull time) increases with the decrease in the duration of the lull time almost as fast as the probability of the operation time does. It is only for very long times of operation that the probability P_l of exceeding a certain time limit starts to play an appreciable role. We conclude, therefore, that in the situation in question the human factor associated with the decision-making times plays a significant role, as far as safe landing is concerned. The developed model enables one to quantitatively assess this role.

Example 3.2. Let, for example, the number of ship oscillations during the time of landing be $n^* = 5$, the required (specified) probability of safe landing be as high as $P_D = 0.9999$, the vertical velocity due to the ship motions during the lull period be $D_{\dot{z}} = 0.030\,\text{m/s}$, and the extreme value of the relative vertical velocity computed as the difference between the specified (allowable) velocity v^* of the helicopter and the actual ground landing velocity v_0, be $v^* - v_0 = 0.8\,\text{m/s}$. Then the level of the relative velocity at the moment of landing is

$$\dot{z}^* = \sqrt{2D_{\dot{z}}\left[\ln n^* - \ln\left(-\ln P_D + \exp\left(-n^*\right)\right)\right]}$$

$$= \sqrt{2 \times 0.030\left[\ln 5 - \ln\left(-\ln 0.9999 + \exp(-5)\right)\right]} = 0.629\,\text{m/s} \triangleleft 0.8\,\text{m/s}.$$

Hence, landing can be allowed and is expected to be safe.

3.9 Conclusions

The following major conclusions can be drawn from the carried out analysis:

- The developed probabilistic model enables one to assess the role of the human factor, along with other uncertainty sources, in the HLS situation. Safe landing can be expected if the probability that it takes place during the lull time is sufficiently high. The developed simple and easy-to-use formulas enable to evaluate this probability.

- The suggested model can be used in the analysis of the landing situation, as well as in the probabilistic assessment of the strength of the helicopter undercarriage. It can be used also when developing guidelines for personnel training, and to determine if, for the specified probability of the safe landing, a human can be trained to an extent that his/her decision-making and reaction times will be short enough for such a specified probability, and, if not, whether special equipment and instrumentation should be developed and installed.

- Plenty of additional risk analyses and human psychology–related effort will be needed, however, to make such guidelines practical.

References

1. W.A. O'Neil, "The Human Element in Shipping," Keynote Address. In *Biennial Symposium of the Seafarers International Research Center*, Cardiff, Wales, June 29, 2001.
2. US Coast Guard.*Prevention Through People Quality Action Team Report*, USCG, Washington, DC, 1995.
3. P. Boisson, *Safety at Sea: Policies, Regulations and International Law*, Bureau Veritas, Paris, 1999.
4. G. Miller, "Human Factor Engineering (HFE): What Is It and How It Can Be Used to Reduce Human Errors in the Offshore Industry (OTC 10876)." In *Proceedings of 1999 Offshore Technology Conference*, Houston, TX, May 3–6, 1999.
5. USCG Headquarters. *Training System Standard Operating Procedures, vol. 2 Analysis*, USCG, Washington, DC, 2008.
6. USCG Research and Development Center. *USCG Guide for the Management of Crew Endurance and Risk Factors* (Report No. CF-D-13-01). USCG, Groton, CT, 2001.
7. D.C. Foyle and B.L. Hooey, *Human Performance Modeling in Aviation*, CRC Press, New York, 2008.
8. E. Suhir, *Applied Probability for Engineers and Scientists*, McGraw-Hill, New York, 1997.

9. US Patent #6064924, "Method and System for Predicting Ship Motion or the Like to Assist in Helicopter Landing," 1970.
10. US Patent #3516375, "Horizon Indicators for Assisting Helicopter Landing on Ships," 1970.
11. L.A. Mcgee, C.H. Paulk, Jr., S.A. Steck, S.F. Schmidt, and A.W. Merz, *Evaluation of the Navigation Performance of Shipboard-VTOL-Landing Guidance Systems*, NASA-Ames Research Center, Mountain View, CA, 1979.
12. A.V. Phatak, M.S. Karmali, and C.H. Paulk, Jr., *Ship Motion Pattern Directed VTOL Letdown Guidance*, NASA Ames Research Center, Mountain View, CA, 1983.
13. C.H. Paulk, Jr. and A.V. Phatak, *Evaluation of a Real-Time Predictive Guidance Law for Landing VTOL Aircraft at Sea*, NASA-Ames Research Center, Mountain View, CA, 1984.
14. J. Blackwell and R.A. Feik, "A Mathematical Model of the On-Deck Helicopter/Ship Dynamic Interface." In *Aerodynamic Technical Memorandum*, ADSB 130078, September 1988. Aeronautical Research Laboratory, Melbourne, VIC, 1988.
15. D.G. Thomson, F.N. Coton, and R.A.M. Galbraith, "Simulator Study of Helicopter Ship Landing Procedures Incorporating Measured Flow Data." In *Proceedings of the Institution of Mechanical Engineers, Part G: Journal of Aerospace Engineering*, vol. 219, No. 5, 2005, pp. 411–427.
16. "Swedish VISBY Class Corvette Sees First Helicopter Deck Landing: Ship-Helicopter Operation Limits," 2006.
17. D. Carico and B. Ferrier, "Evaluating Landing Aids to Support Helicopter/Ship Testing and Operations." In *IEEE Aerospace Conference*, 2006, vol. 4, No. 4–11, March 2006.
18. E. Suhir, "Adequate Underkeel Clearance (UKC) for a Ship Passing a Shallow Waterway: Application of the Extreme Value Distribution (EVD)." In *Proceedings of the OMAE 2001*, Rio-de-Janeiro, Brazil, OMAE2001/S&R-2113, 2001.
19. T.H. Soukissian and P.E. Samalikos, "Analysis of the Duration and Intensity of the Sea State Using Segmentation of Significant Wave Height Time Series." In *16th International Offshore and Polar Engineering Conference*, San Francisco, CA, May 28–June 2, 2006.

4

Fundamentals of Probabilistic Aerospace Electronics Reliability Engineering

4.1 Today's Practices: Some Problems Envisioned and Questions Asked

- Electronic products that underwent highly accelerated life testing (HALT) [1], passed the existing qualification tests (QTs) and survived burn-in testing (BIT) often exhibit nonetheless premature field failures. Are these methodologies and practices, and particularly the accelerated test (AT) procedures, adequate [2]? Do electronic industries need new approaches to qualify their products, and if they do, what could and should be done differently? Could the existing practices be improved to an extent that if the product passed the reliability tests, there is a way to assure that it will satisfactorily (i.e., failure free) perform in the field?

- In many applications, such as, for example, aerospace, military, long-haul communications, medical, and others, high reliability of electronics materials and products is particularly imperative. Could the operational (field) reliability of an electronic product be assured, if it is not predicted (i.e., not quantified)? And if such a quantification is found to be necessary, could that be done on the deterministic (i.e., on a nonprobabilistic) basis [3–5]?

- Should electronic product manufacturers keep shooting for an unpredictable and, perhaps, unachievable very long, such as, for example, 20 years or so, product lifetime or, considering that every 5 years a new generation of devices appears on the market, should the manufacturers settle for a shorter, but substantiated, predictable, and assured lifetime, with an assessable high probability of nonfailure? And how should such a lifetime be related to the acceptable (specified) probability of nonfailure for a particular product and application?

73

- Considering that the principle of superposition does not work in reliability engineering, how should one establish the list of the crucial accelerated tests (AT)s, the adequate physically meaningful and realistic stressors, and their combinations and levels?

- The best engineering product is, as is known, the best compromise between the requirements for its reliability, cost effectiveness, and short-as-possible time-to-market (to completion); it goes without saying that, in order to make optimization possible, the reliability of such a product has to be quantified, but how could one do that?

- The bathtub curve, the experimental "reliability passport" of a mass-fabricated product, reflects the inputs of two critical irreversible processes—the statistics-of-failure process that results in a reduced failure rate with time (this is particularly evident from the infant mortality portion of the curve) and the physics-of-failure (aging, degradation) process that leads to an increased failure rate with time (this trend is explicitly exhibited by the wear out portion of the bathtub diagram). Could these two critical processed be separated [6]? The need for that is due to the obvious incentive to minimize the role and the rate of aging, and this incentive is especially significant for products like lasers, solder joint interconnections, and others, which are characterized by long wear out portions and when it is economically infeasible to restrict the product's lifetime to the steady-state situation, when the two irreversible processes in question compensate each other to a greater or lesser extent.

- A related question has to do with the fact that real time degradation is a very slow process. Could physically meaningful and cost-effective methodologies for measuring and predicting the degradation (aging) rates and consequences be developed? In the outline that follows, some of the above problems are addressed with an objective to show how the recently suggested probabilistic design for reliability (PDfR) concept can be effectively employed for making a viable electronic or an optical device into a reliable and marketable product.

4.2 Accelerated Testing

Shortening of electronic product's design and development time does not allow in today's industrial environment for time-consuming reliability investigations. To get maximum reliability information in minimum time and at minimum cost is the major goal of an electronic or an optical product manufacturer. It is also impractical to wait for failures, when the lifetime of a typical today's electronic or optical product is hundreds of thousands of hours, regardless of whether it could or could not be predicted with sufficient accuracy. Accelerated testing is, therefore, both a must and a powerful means in electronics manufacturing. Different types of such testing are shown and their features are briefly indicated in Table 4.1[7].

A typical example of product development testing (PDT) is shear-off testing conducted when there is a need to determine the most feasible bonding material and its thickness, and/or to assess its strength and/or evaluate the shear modulus of this material.

HALT is currently widely employed, in different modifications, with an intent to determine the product's reliability weaknesses, assess reliability limits, ruggedize the product by applying elevated stresses (not necessarily mechanical and not necessarily limited to the anticipated field stresses) that could cause field failures, and provide large (although, actually, unknown) safety margins over expected in-use conditions. HALT often involves stepwise stressing, rapid thermal transitions, and other means that enable one to carry out testing in a time- and cost-effective fashion. HALT is sometimes referred to as a "discovery" test. It is not a qualification test though (i.e., not a "pass/fail" test). It is the qualification testing that is the major means for making a viable electronic device into a reliable marketable product. While many HALT aspects are different for different manufacturers and often kept as proprietary information, qualification tests and standards are the same for the given industry and product type.

Qualification tests (QT) are the major means to make a viable device into a marketable product.

Burn-in testing (BIT) is post-manufacturing testing. Mass fabrication, no matter how good the design concepts and the fabrication technologies are, inevitably generates, in addition to desirable and robust ("strong") products, also some undesirable and unreliable ("weak") devices ("freaks"), which, if shipped to the customer, will most likely fail in the field. BIT is supposed to detect and to eliminate such "freaks." As a result, the final bathtub curve of a product that underwent burn-in is not supposed to contain the infant mortality portion. In today's practice a successful burn-in, a destructive test for the "freaks" and a nondestructive test for the healthy devices, is often run within the framework of, and concurrently with, HALT.

But are today's practices based on the above reliability testing always adequate? The funny, but quite practical, definition of a sufficiently robust product is that, as some reliability managers put it, "reliability it is when the customer comes back, not the product." It is well known, however, that electronic and optical products that underwent HALT, passed the existing QTs, and survived BITs often exhibit premature operational failures. Are the existing practices adequate?

Many reliability engineers think that one crucial shortcoming of today's reliability assurance practices is that they are not based on the sufficient understanding of the underlying reliability physics for the particular product, its time in operation, and operation conditions. But how could one

TABLE 4.1

Accelerated Test Categories

Accelerated Testing Type	Product Development Testing	Highly Accelerated Life Testing (HALT)	Qualification Testing	Burn-In Testing	Failure-Oriented-Accelerated Testing (FOAT)
Objective	Technical feedback to assure that the taken design approach is acceptable	Ruggedize the product and assess its reliability limits	Proof that the product meets the existing reliability standard(s)	Eliminate the infant mortality part of the bathtub curve	Understanding the physics of failure, checking the applicability of the accepted failure model, assessing the probability of failure
End point	Type, time, level, or the number of the observed failures	Predetermined number or percentage of failures	Predetermined time and/or the number of cycles and/or excessive (unexpected) number of failures	Predetermined time and/or loading level	Predetermined number or percentage (typically 50%) of failure
Follow-up activity	Failure analysis, design decision	Failure analysis	Pass/fail decision	Shipping of sound devices	Probabilistic analysis of the test data
Perfect test	Specific definitions	No failures in a long time			Numerous failures in a short time

understand the physics of failure without conducting a highly focused and highly cost-effective failure-oriented-accelerated testing (FOAT)? If such testing is considered, it should be geared to a particular adequate, simple, easy-to-use, and physically meaningful predictive model.

Predictive modeling has proven to be a highly useful means for understanding the physics of failure and designing the most practical accelerated tests in electronic and photonic engineering. It has been recently suggested that FOAT should be considered as the experimental basis of the fruitful flexible and physically meaningful approach PDfR of electronic and optical products. This approach is based on the following 10 major requirements ("commandments") reflecting the rationale behind the PDfR concept. In a way, these principles ("commandments") are similar to the requirements for the probabilistic HITL-related situations in aerospace ergonomics.

4.3 PDfR and Its Major Principles ("10 Commandments")

The PDfR concept is an effective means for improving the state-of-the-art in the electronics and photonics reliability field by quantifying, on the probabilistic basis, the operational reliability of a material or a product by predicting the probability of its likely failure under the given loading conditions and after the given service time, and to use this probability as a suitable and physically meaningful criterion of the expected product's performance in the field. The following 10 major (governing) principles ("commandments") reflect the rationale behind the PDfR concept:

1. When reliability of a product is imperative, the ability to predict it is a must; it could be argued that operational reliability cannot be assured, if it is not quantified at the design stage.

2. Nothing is perfect; in effect, the difference between a highly reliable and an insufficiently reliable product is "merely" in the level of their never-zero probability of failure, and therefore, such a quantification should be done on the probabilistic basis.

3. Reliability evaluations cannot be delayed until the product is made and should start at the design stage and should be taken care of at all the significant stages of the product's lifetime: at the design stage, when reliability is conceived; at the accelerated testing stage, using electrical, optical, environmental, and mechanical instrumentation to monitor the product's behavior; at the production and manufacturing stage, when reliability is actually implemented; and, if necessary, appropriate, and possible, reliability should be maintained in

the field during the product's operation; then there will be a reason to believe that a "genetically healthy" product is created and its "health" could be maintained by using various popular prognostics and health monitoring (PHM) methods, as well as redundancy, trouble-shooting, and other more or less important and established means that could be considered to maintain adequate reliability levels, especially if the "genetic health" of the product is not very high.

4. A product's reliability cannot be low, of course, but need not be higher than necessary either: it has to be adequate for the given product and application, considering the projected product's lifetime, environmental conditions, consequences of failure, manufacturing and repair costs, etc.

5. The best product is the best compromise between the requirements for its reliability, cost effectiveness, and time to market; it goes without saying that such a compromise cannot be addressed and achieved if reliability is not quantified.

6. As follows from the above considerations, one cannot design a product with quantified, optimized, and assured reliability by limiting the effort to the widely used today "black box"—HALT; understanding the underlying physics of failure is crucial, and therefore highly cost-effective and highly focused FOAT should be always considered and conducted as a possible and natural extension of HALT.

7. FOAT, unlike HALT, is a "white/transparent box" aimed, first of all, at understanding the physics of failure; FOAT should be geared to a limited number of predetermined simple, easy-to-use, and physically meaningful predictive reliability models and should be viewed as the "reliable" experimental basis and important constituent part of the PDfR concept.

8. The recently suggested physically meaningful, powerful, easy-to-use, and flexible multiparametric Boltzmann–Arrhenius–Zhurkov (BAZ) model can be used as a suitable one for the assessment of the remaining "useful" life (RUL) of an electronic or a photonic product.

9. Predictive modeling, not limited to FOAT models, is a powerful means to carry out, if necessary, sensitivity analyses (SAs) with an objective to quantify and practically nearly eliminate failures by making the probability of failure sufficiently low; this principle could be referred to as the "principle of practical confidence."

10. Consideration of the role of the human factor is highly desirable and can be included in the PDfR effort: not only "nothing" but also "nobody" is perfect, and the human role in assessing the likelihood of the adequate performance of a product or an outcome of a mission is critical and can and often should be considered.

4.4 FOAT ("Transparent Box") as an Extension of HALT ("Black Box")

As has been indicated, a highly focused and highly cost-effective FOAT is the experimental foundation and the "heart" of the PDfR concept. FOAT should be conducted in addition to and, in some cases, even instead of HALT, especially for new products, whose operational reliability is unclear and for which no experience is accumulated and no best practices or HALT methodologies are yet developed and established. Predictions, based on the FOAT and subsequent probabilistic predictive modeling, might not be perfect, at least at the beginning, but it is still better to pursue such an effort rather than to turn a blind eye on the fact that there is always a nonzero probability of the product's failure. Understanding the underlying reliability physics for the product performance is critical. If one sets out to understand the physics of failure in an attempt to create a failure-free product (in accordance with the "principle of practical confidence"), conducting a FOAT type of an experiment is imperative. FOAT's objective is to confirm the usage of a particular more or less well-established predictive model, to confirm (say, after HALT is conducted) the physics of failure, and to establish the numerical characteristics (activation energy, time constant, sensitivity factors, etc.) of the particular FOAT model of interest.

FOAT could be viewed as an extension of HALT. While HALT is a "black box" (i.e., a methodology that can be perceived in terms of its inputs and outputs without a clear knowledge of the underlying physics and the likelihood of failure), FOAT, on the other hand, is a "transparent box," whose main objective is to confirm the use of a particular reliability model that reflects a specific anticipated failure mode. The major assumption is, of course, that this model should be valid in both AT and in actual operation conditions.

HALT does not measure (does not quantify) reliability. FOAT does. HALT can be used for "rough tuning" of product reliability, and FOAT should be employed when "fine tuning" is needed (i.e., when there is a need to quantify, assure, and even specify the operational reliability of a product). HALT tries to "kill many unknown birds with one (also not very well known) stone." HALT has demonstrated, however, over the years its ability to improve robustness through a "test-fail-fix" process, in which the applied stresses (stimuli) are somewhat above the specified operating limits. This "somewhat above" is based, however, on an intuition, rather than on a calculation. There is a general perception that HALT might be able to quickly precipitate and identify failures of different origins.

FOAT and HALT could be carried out separately, or might be partially combined in a particular AT effort. Since the principle of superposition does not work in reliability engineering, both HALT and FOAT use, when appropriate, combined stressing under various anticipated stimuli (stressors).

New products present natural reliability concerns, as well as significant challenges at all the stages of their design, manufacture, and use. An appropriate combination of HALT and FOAT efforts could be especially useful for ruggedizing and quantifying reliability of such products. It is always necessary to correctly identify the expected failure modes and mechanisms, and to establish the appropriate stress limits of HALTs and FOATs with an objective to prevent "shifts" in the dominant failure mechanisms. There are many ways of how this could be done (see, e.g., [8]). The FOAT-based approach could be viewed as a quantified and reliability physics-oriented HALT. The FOAT approach should be geared to a particular technology and application, with consideration of the most likely stressors.

4.5 Design for Reliability of Electronics Systems: Deterministic and Probabilistic Approaches

Design for reliability (DfR) is, as is known, a set of approaches, methods, and best practices that are supposed to be used at the design stage of an electronic or a photonic product to minimize the risk that the product might not meet the reliability objectives and customer expectations. When a deterministic approach is used, reliability of a product could be based on the belief that a sufficient reliability level will be assured if a high enough safety factor (SF) is used. The deterministic SF is defined as the ratio $SF = \dfrac{C}{D}$ of the capacity ("strength") C of the product to the demand ("stress') D. The PDfR SF is introduced, as the ratio of the mean value $\prec \psi \succ$ of the safety margin $SM = \Psi = C - D$ to its standard deviation $\hat{s}$, so that the probabilistic safety factor is $SF = \dfrac{\prec \psi \succ}{\hat{s}}$. When the random time-to-failure (TTF) is of interest, the SF can be found as the ratio of the mean time to failure (MTTF) to the standard deviation (STD) of the TTF. The use of SF as a measure of the probability of failure (PoF) is more convenient than the direct use of the PoF itself. This is because this probability is expressed, for highly reliable and, hence, typical electronic or optical products, by a number, which is very close to one, and, for this reason, even significant changes in the product's design, with an appreciable impact on its reliability, might have a minor effect on the level of the PoF, at least the way it appears to and is perceived by the user. The SF tends to infinity, when the probability of nonfailure tends to one. The PoF (the level of the SF) should be chosen depending on the experience, anticipated operation conditions, possible consequences of failure, acceptable risks, available and trustworthy information about the capacity and the demand, accuracy with which the capacity and the demand are determined, possible costs

Probabilistic Aerospace Electronics Reliability Engineering 81

and social benefits, information on the variability of materials and structural parameters, fabrication technologies and procedures, and so on.

4.6 Two Simple PDfR Models

Example 4.1. The probability of nonfailure of a heat sink experiencing elevated temperature can be found using the Arrhenius equation

$$\tau = \tau_0 \exp\left(\frac{U_0}{kT}\right) \tag{4.1}$$

and the exponential law of reliability, so that

$$P = \exp\left[-\frac{t}{\tau_9} \exp\left(\frac{U_0}{kT}\right)\right] \tag{4.2}$$

Solving this equation for the absolute temperature T, we have

$$T = \frac{U_0/k}{\ln\left(-\frac{\tau_0}{t}\ln P\right)} \tag{4.3}$$

Addressing, for example, surface charge accumulation failure, for which the ratio of the activation energy to the Boltzmann's constant is $\frac{U}{k} = 11,600°K$, assuming that the FOAT-predicted time factor τ_0 is $\tau_0 = 2 \times 10^{-5}$ h, that the customer requires that the probability of failure at the end of the device's service time of $t = 40,000$ h is only $Q = 10^{-5}$, the formula (4.3) yields: $T = 352.3°K = 79.3°C$. Thus, the heat sink should be designed accordingly, and the vendor should be able to deliver such a heat sink.

Example 4.2. The maximum interfacial shearing stress in a solder glass layer in a ceramic electronic package can be computed by the formula:

$$\tau_{max} = kh_g\sigma_{max}. \tag{4.4}$$

Here,

$$k = \sqrt{\frac{\lambda}{\kappa}} \tag{4.5}$$

is the parameter of the interfacial shearing stress;

$$\lambda = \frac{1 - v_c}{E_c h_c} + \frac{1 - v_g}{E_g h_g} \tag{4.6}$$

is the axial compliance of the assembly that consists of the ceramic ("c") and the glass ("g") materials;

$$\kappa = \frac{h_c}{3G_c} + \frac{h_g}{3G_g} \tag{4.7}$$

is its interfacial compliance of this assembly;

$$G_c = \frac{E_c}{2(1+v_c)}, \quad G_g = \frac{E_g}{2(1+v_g)} \tag{4.8}$$

are the shear moduli of the ceramics and the glass materials;

$$\sigma_{max} = \frac{\Delta\alpha\Delta t}{\lambda h_g} \tag{4.9}$$

is the maximum normal stress in the midportion of the glass layer; Δt is the change in temperature from the high soldering temperature to the low (room or testing) temperature, at which the thermal stresses are the highest; $\Delta\alpha = \bar{\alpha}_c - \bar{\alpha}_g$ is the difference in the effective coefficients of thermal expansion (CTEs) of the ceramics and the glass;

$$\bar{\alpha}_{c,g} = \frac{1}{\Delta t} \int_t^{t_0} \alpha_{c,g}(t)dt \tag{4.10}$$

are these coefficients for the given temperature t; t_0 is the annealing temperature; and $\alpha_{c,g}(t)$ are the time-dependent CTEs for the materials in question. In an approximate analysis, one could assume that the axial compliance λ of the assembly is due to the glass layer only, so that

$$\lambda \approx \frac{1 - v_g}{E_g h_g} \tag{4.11}$$

and, therefore, the maximum normal stress in the cross sections of the solder glass layer can be evaluated as

$$\sigma_{max} = \frac{E_g}{1 - v_g} \Delta\alpha\Delta t. \tag{4.12}$$

While the geometric characteristics of the assembly, the change in temperature, and the elastic constants of the materials can be determined with high accuracy, this is not the case for the difference in the CTEs of the brittle materials of the glass and the ceramics. In addition, because of the obvious incentive to minimize this difference, such a mismatch is characterized by a small difference of close and quite appreciable numbers. This circumstance contributes to the uncertainty of the problem in question and justifies the application of the probabilistic approach.

Probabilistic Aerospace Electronics Reliability Engineering

Treating the CTEs of the two materials as normally distributed random variables, we evaluate the probability P that the thermal interfacial shearing stress is compressive (negative) and, in addition, is sufficiently low (i.e., does not exceed an allowable level) [9]. This stress is proportional to the normal stress in the glass layer, which is, in its turn, proportional to the difference $\Psi = \alpha_c - \alpha_g$ of the CTE of the ceramics and the glass materials. One wants to make sure that the requirement

$$0 \le \Psi \le \Psi_* = \frac{\sigma_\alpha}{E_g} \frac{1 - v_g}{\Delta t} \tag{4.13}$$

takes place with a high enough probability.

For normally distributed random variables α_c and α_g, the random variable Ψ is also normally distributed. Its mean value and standard deviation are

$$\prec \psi \succ = \prec \alpha_c \succ - \prec \alpha_g \succ \tag{4.14}$$

and

$$\sqrt{D_\psi} = \sqrt{D_c + D_g}, \tag{4.15}$$

where $\prec \alpha_c \succ$ and $\prec \alpha_g \succ$ are the mean values of the materials' CTEs, and D_c and D_g are the CTE variances. The probability that the condition (4.13) takes place is

$$P = \int_0^{\psi_*} f_\psi(\psi) d\psi = \Phi_1(\gamma * - \gamma) - \left[1 - \Phi(\gamma)\right], \tag{4.16}$$

where

$$\Phi(t) = \frac{1}{\sqrt{2\pi}} \int_{-\infty}^{t} e^{-t^2/2} dt \tag{4.17}$$

is the Laplace function (probability integral),

$$\gamma = \frac{\prec \psi \succ}{\sqrt{D_\psi}} \tag{4.18}$$

is the SF for the CTE difference, and

$$\gamma^* = \frac{\psi^*}{\sqrt{D_\psi}} \tag{4.19}$$

is the SF for the acceptable level of the allowable stress.

If, for example, the elastic constants of the solder glass are $E_g = 0.66 \times 10^6 \, \text{kg/cm}^2$ and $v_g = 0.27$, the sealing (fabrication) temperature is 485°C, the lowest (testing) temperature is −65°C (so that $\Delta t = 550$°C), the computed effective CTEs at this temperature are $\bar{\alpha}_g = 6.75 \times 10^{-6} 1/$°C and $\bar{\alpha}_c = 7.20 \times 10^{-6} 1/$°C, the standard deviations of these STEs are $\sqrt{D_c} = \sqrt{D_g} = 0.25 \times 10^{-6} 1/$°C, respectively, and the (experimentally obtained) ultimate compressive strength for the solder glass material is $\sigma_u = 5500 \, \text{kg/cm}^2$. With the acceptable SF of, say, 4, we have $\sigma^* = \sigma_u/4 = 1375 \, \text{kg/cm}^2$. The allowable level of the parameter ψ is, therefore,

$$\psi_* = \frac{\sigma_a}{E_g} \frac{1 - v_g}{\Delta t} = \frac{1375}{0.66 \times 10^6} \frac{0.73}{550} = 2.765 \times 10^{-6} 1/°C.$$

The mean value $\prec \psi \succ$ and variance D_ψ of this parameter are

$$\prec \psi \succ = \prec \alpha_c \succ - \prec \alpha_g \succ = 0.45 \times 10^{-6} 1/°C$$

and

$$D_\psi = D_c + D_g = 0.25 \times 10^{-12} \left(1/°C \right)^2,$$

respectively. Then the predicted SFs are $\gamma = 1.2726$ and $\gamma^* = 7.8201$, and the corresponding probability of nonfailure of the seal glass material is

$$P = \Phi_1(\gamma^* - \gamma) - \left[1 - \Phi_1(\gamma) \right] = 0.898.$$

If the standard deviations of the materials CTEs were only $\sqrt{D_c} = \sqrt{D_g} = 0.1 \times 10^{-6} 1/$°C, then the SFs would be much higher: $\gamma = 3.1825$ and $\gamma^* = 19.5559$, and the probability of nonfailure would be as high as $P = 0.999$.

4.7 BAZ Model: Possible Way to Quantify Reliability

The Arrhenius model (4.1) can be generalized, in the presence of an external stress σ, as follows:

$$\tau = \tau_0 \exp\left(\frac{U_0 - \gamma\sigma}{kT} \right) \tag{4.20}$$

[10]. This is Boltzmann–Arrhenius–Zhurkov (BAZ) model. It can be used when the material experiences the combined action of elevated temperature T and external loading σ. Although in Zhurkov's fracture mechanics tests the loading σ was always a constant mechanical tensile stress, it has been

recently suggested that any other stimulus of importance (voltage, current, thermal stress, humidity, radiation, random vibrations, etc.) can be used as such a stress. The effective activation energy

$$U = kT \ln \frac{\tau}{\tau_0} = U_0 - \gamma\sigma \tag{4.21}$$

plays in the BAZ model the role of the stress-free energy U_0 in the Arrhenius model ($\sigma = 0$). The BAZ model can be obtained as the steady-state solution to the Fokker–Planck equation in the theory of Markovian processes and represents the worst-case scenario,so that the reliability predictions based on the steady-state BAZ model are conservative and therefore advisable in engineering practice [11].

Let the lifetime τ in the BAZ model be viewed as the MTTF. Such an assumption suggests that if the exponential law of probability $P = \exp(-\lambda t)$ of nonfailure is used, the MTTF corresponds to the moment of time when the entropy of this law reaches its maximum value. Indeed, from the formula $H(P) = -P \ln P$, we obtain that the maximum value of the entropy $H(P)$ is equal to e^{-1} and takes place for $P = e^{-1.} = 0.3679$. This yields: $t = -\frac{1}{\lambda} \ln P = -\tau \ln P = \tau$. Thus, the MTTF in the Arrhenius and BAZ equations is, in effect, the time when the enthropy of the process $P = P(t)$ is the largest.

4.8 Multiparametric BAZ Model

Let us elaborate on the substance of the multiparametric BAZ model [12] using as an example a situation when the product of interest is subjected to the combined action of the elevated relative humidity H and elevated voltage V. Let us assume that the failure rate of a product is determined by the level of the measured leakage current: $\lambda = \gamma_1 I$. Then one can seek the probability of the product's nonfailure as

$$P = \exp\left[-\gamma_1 It \exp\left(-\frac{U_0 - \gamma_H H - \gamma_V V}{kT}\right)\right]. \tag{4.22}$$

Here the γ factors reflect the sensitivities of the device to the change in the corresponding stressors. Although only two stressors are selected here—the relative humidity H and the elevated voltage V—the model can be easily made multiparametric (i.e., generalized for as many stimuli as necessary). The sensitivity factors γ should be determined from the FOAT when the combined action of all the stimuli (stressors) of importance is considered. The physical meaning of the distribution (4.22) could be seen from the formulas

$$\frac{\partial P}{\partial I} = -\frac{H(P)}{I}, \quad \frac{\partial P}{\partial I} = -\frac{H(P)}{\partial t}, \quad \frac{\partial P}{\partial U_0} = \frac{H(P)}{kT}, \tag{4.23}$$

$$\frac{\partial P}{\partial H} = -\frac{H(P)}{kT}\gamma_H = -\gamma_H \frac{\partial P}{\partial U_0}, \quad \frac{\partial P}{\partial V} = -\frac{H(P)}{kT}\gamma_V = -\gamma_V \frac{\partial P}{\partial U_0},$$

where $H(P) = -P\ln P$ is the entropy of the probability $P = P(t)$ of nonfailure. The following conclusions can be made based on these formulas:

1. The change in the probability of nonfailure always increases with an increase in the entropy (uncertainty) of the distribution and decreases with an increase in the leakage current and with time. This makes certainly physical sense.

2. The last two formulas in (4.23) show the physical meaning of the sensitivity factors γ: these factors can be found as the ratios of the change in the probability of nonfailure with respect to the corresponding stimuli to the change of this probability with the change in the stress-free activation energy.

The equation for the probability of nonfailure contains four empirical parameters: the stress-free activation energy U_0 and three sensitivity factors γ: leakage current factor, relative humidity factor, and elevated voltage factor. Here is how these factors could be obtained from the FOAT data.

First one should run the FOAT for two different temperatures T_1 and T_2, keeping the levels of the relative humidity H and elevated voltage V the same in both tests; recording the percentages (values) P_1 and P_2 of nonfailed samples (or values $Q = 1 - P_1$ and $Q = 1 - P_2$ of the failed samples); assuming a certain criterion of failure (say, when the level of the measured leakage current exceeds a certain value I_*). Then the following two relationships can be obtained:

$$P_1 = \exp\left[-\gamma_I I_* t_1 \exp\left(-\frac{U_0 - \gamma_H H - \gamma_V V}{kT_1}\right)\right],$$

$$P_2 = \exp\left[-\gamma_I I_* t_2 \exp\left(-\frac{U_0 - \gamma_H H - \gamma_V V}{kT_2}\right)\right]. \tag{4.24}$$

Since the numerators in these relationships are the same, the following equation must be fulfilled for the sought sensitivity factor γ_I of the leakage current:

$$f(\gamma_I) = \ln\left(\frac{\ln P_1}{I_* t_1 \gamma_I}\right) - \frac{T_1}{T_1}\ln\left(\frac{\ln P_2}{I_* t_2 \gamma_I}\right) = 0. \tag{4.25}$$

Probabilistic Aerospace Electronics Reliability Engineering 87

Here t_1 and t_2 are the times at which the failures were detected. It is expected that more than just two series of FOAT and at more than two temperature levels are conducted, so that the sensitivity parameter γ_I could be determined with a high enough accuracy.

At the second step, FOAT at two relative humidity levels H_1 and H_2 should be conducted for the same temperature and voltage. This leads to the following relationship:

$$\gamma_H = \frac{kT}{H_1 - H_2}\left[\ln\left(\frac{\ln P_1}{I*t_1\gamma_I}\right) - \ln\left(-\frac{\ln P_2}{I*t_2\gamma_I}\right)\right]. \tag{4.26}$$

Similarly, at the *next step* of FOAT, by changing the voltages V_1 and V_2, the following expression for the sensitivity factor γ_V can be obtained:

$$\gamma_V = \frac{kT}{V_1 - V_2}\left[\ln\left(-\frac{\ln P_1}{I*t_1\gamma_I}\right) - \ln\left(\frac{\ln P_2}{I*t_2\gamma_I}\right)\right]. \tag{4.27}$$

Finally, the stress-free activation energy can be computed as

$$U_0 = \gamma_H H + \gamma_V V - kT\ln\left(-\frac{\ln P}{I*t\gamma_I}\right) \tag{4.28}$$

for any consistent humidity, voltage, temperature, and time. The above relationships could be obtained particularly also for the case of zero voltage. This will provide additional information of the materials and device reliability characteristics.

4.9 The Total Cost of Reliability Could Be Minimized: Elementary Example

Let us show [13], using an elementary modeling-based example, how cost optimization could be, in principle, done. Let us assume that the cost of achieving and improving reliability can be estimates based on the formula

$$C_R = C_R(0)\exp\left[r(R - R_0)\right], \tag{4.29}$$

where $R = MTTF$ is the actual level of the MTTF, R_0 is the specified MTTF level, $C_R(0)$ is the cost of achieving the R_0 level of reliability, and r is the cost factor associated with reliability improvements. Similarly, let us assume that the cost of reliability repair can also be assessed by a similar formula

$$C_F = C_F(0)\exp\left[-f(R-R_0)\right], \tag{4.30}$$

where $C_F(0)$ is the cost of restoring the product's reliability, and f is the factor of the reliability restoration (repair) cost. The latter formula reflects a natural assumption that the cost of repair is lower for a product of higher reliability. The total cost $C = C_R + C_F$ has its minimum

$$C_{min} = C_R\left(1+\frac{r}{f}\right) = C_F\left(1+\frac{f}{r}\right), \tag{4.31}$$

when the minimization condition $rC_R = fC_F$ is fulfilled. Let us further assume that the factor r of the reliability improvement cost is inversely proportional to the MTTF (dependability criterion), and the factor f of the reliability restoration cost is inversely proportional to the mean time to repair MTTR (reparability criterion). Then the minimum total cost is

$$C_{min} = \frac{C_R}{K} = \frac{C_F}{1-K}, \tag{4.32}$$

where the availability K (i.e., the probability that the product is sound and is available to the user any time at the steady-state operations) is expressed as

$$K = \frac{1}{1+\dfrac{\prec t_r \succ}{\prec t_f \succ}} = \frac{1}{1+\dfrac{MTTR}{MTTF}}. \tag{4.33}$$

In this formula $\prec t_f \succ\ = MTTF$ is the mean TTF, and $\prec t_r \succ\ = MTTR$ is the mean time to repair. The above result obtained for the total minimum cost establishes, in an elementary way, the relationship between the minimum total cost of achieving and maintaining (restoring) the adequate reliability level and the availability criterion. The obtained relationship quantifies the intuitively obvious fact that the total cost of the product depends on both the total cost and the availability of the product. The formula

$$\frac{C_F}{C_R} = \frac{1}{K} - 1 \tag{4.34}$$

that follows from the above derivation indicates that if the availability index K is high, the ratio of the cost of repairs to the cost aimed at improved reliability is low. When the availability index is low, this ratio is high. Again, this intuitively obvious result is quantified by the obtained simple relationship. The above reasoning can be used, particularly, to interpret the availability index from the cost-effectiveness point of view: the index

$$K = \frac{C_R}{C_{min}}$$

reflects, in effect, the ratio of the cost of improving reliability to the minimum total cost of the product associated with its reliability level.This and similar, even elementary, models can be of help, particularly, when there is a need to minimize costs without compromising reliability (i.e., in various optimization analyses).

4.10 Possible Next Generation of the Qualification Tests (QTs)

The next-generation QT could be viewed as a "quasi-FOAT," "mini-FOAT," sort of an "initial stage of FOAT" that more or less adequately replicates the initial nondestructive, yet full-scale, stage of FOAT. The duration and conditions of such a "mini-FOAT" QT could and should be established based on the observed and recorded results of the actual FOAT, and should be limited to the stage when no failures, or a predetermined and acceptable small number of failures in the actual full-scale FOAT, were observed. Various prognostics and health monitoring (PHM) technologies ("canaries") could be concurrently employed to make sure that the safe limit is established correctly. FOAT should be thoroughly designed, implemented, and analyzed, so that the QT is based on the trustworthy FOAT data.

4.11 Physics-of-Failure BAZ Model Sandwiched between Two Statistical Models: Three-Step Concept

4.11.1 Incentive/Motivation

When encountering a particular reliability problem at the design, fabrication, testing, or an operation stage of a product's life, and considering the use of predictive modeling to assess the seriousness and the likely consequences of a detected failure, one has to choose whether a statistical, or a physics-of-failure-based, or a suitable combination of these two major modeling tools should be employed to address the problem of interest and to decide on how to proceed. A *three-step concept* (TSC) has been recently suggested as a possible way to go in such a situation. The Bayes formula can be used at the first step as a technical diagnostics tool, with an objective to identify, on the probabilistic basis, the faulty (malfunctioning) device or devices from the obtained signals ("symptoms of faults"). The BAZ model and particularly its multiparametric extension can be employed at the second step to assess the RUL of the faulty device or devices. If the RUL is still long enough, no action might be needed; if it is not, corrective restoration action becomes necessary. In any event, after the first two steps are carried out, the device is put back into operation, provided that the assessed probability of its continuing failure-free operation is found to be satisfactory. If an operational failure nonetheless occurs, the third step should be undertaken to update reliability. Statistical beta-distribution, in which the probability of failure is treated as a random variable, is suggested to be used at this step. While various statistical methods and approaches, including Bayes formula and beta-distribution, are well known and widely used in numerous applications for many decades, the BAZ model was introduced only several years ago. The suggested concept is illustrated by a numerical example geared to the use of the PHM effort applied, for example, in en-route flight mission.

4.11.2 Background

The art and science of reliability engineering employs both statistical and probabilistic modeling approaches. Although these approaches are closely related, they pursue different objectives and usually have different times of application in the lifetime of a product. While statistical approaches are *a posteriori* ones, probabilistic predictive modeling is an *a priori* approach, and, as such, is naturally applied first at the design stage of the product's life. Reliability is conceived at the design stage of a product and should be taken care of, first of all, at this stage. If, owing to the design for reliability (DfR) [1–6] and, particularly, the PDfR [7–9] efforts, one manages to create a "genetically healthy" product, then the powerful technical diagnostics "checkups" [10,14] and subsequent "therapeutic" PHM treatments [11–13,15–31] at later stages of the product's life will be dramatically facilitated and will have much better chances to succeed. If such a combined health management activity takes place and is successful, then there is a reason to believe that, by analogy with human's health, a durable and failure-free service life of an aerospace electronic or a photonic product could be achieved, managed, and assured.

In the analysis that follows, a TSC is suggested as a possible PDfR-type probabilistic predictive modeling methodology. This methodology could be considered, whenever appropriate and possible, but not necessarily implemented, when there is a need to obtain, on the continuous basis and at different stages of the product's life, the most trustworthy and comprehensive information, affecting its reliability. The statistical Bayes formula [32–39] is used at the TSC first step as a technical diagnostics tool, with an objective to identify the faulty (malfunctioning) device or devices from the obtained signals ("symptoms of faults"). The physics-of-failure-based BAZ and particularly multiparametric BAZ models [40–44] can be employed at the TSC second step. The objective of this step is to assess the RUL of the malfunctioning device or devices. Statistical beta-distribution [45–47] is suggested to be employed as the possible third step. This step should be undertaken when there is a need to update reliability, if failure occurs despite the expected low probability of failure. Although each of the three major constitutive and subsequent TSC steps—Bayes' formula-based technical diagnostics effort, BAZ-based RUL assessment and, beta-distribution-based reliability update—can be and, in effect, have been employed separately in various particular reliability related tasks, in the TSC methodology addressed here, these parts are combined in such a way that the output of (information from) the previous step is viewed and used as input information for the subsequent step.

The TSC is a rather general approach (i.e., is not specific to a particular failure mechanism or loading condition). The approach could be applied at an assembly, device (component), subsystem, or system level and helps the decision maker to establish the most feasible next step to go at different stages of the product's life. Various statistical methods and approaches, and particularly Bayes' formula and beta-distribution, are well known and have been widely used in numerous applications for many years. The BAZ and the multiparametric BAZ models have been introduced, however, in the microelectronic reliability field as an effective physics-of-failure-based PDfR tool only several years ago. Their attributes are addressed in this analysis therefore in greater detail than the Bayes formula and beta-distribution.

It is noteworthy that while some PDfR models were applied in the past to particular electronics and photonics systems [48–51], it has been only recently suggested [7,8,52] that the PDfR approach be used in microelectronic reliability engineering, as an effective reliability assurance tool, on a wide scale. Highly focused and highly cost-effective FOAT [53] followed by physically meaningful and simple predictive modeling and extensive SAs are the three major constituents of the PDfR concept. FOAT could be viewed as an extension of the currently widely used HALT [52]. As to the multiparametric BAZ models [41,42], they are the most flexible, comprehensive, and well-substantiated ones in the PDfR effort: many models of the FOAT type (Eyring's equation for the mechanical stress, Peck's equation for relative humidity, Black's equation for current density, etc.) contain Arrhenius equation, a special type of the BAZ model, as their core part. If appropriate and desirable, the BAZ and multiparametric BAZ models can include many other advanced probabilistic methods, such as, for example, extreme value distribution, Weibull distribution, normal or log-normal distributions, and so on [10,42].

The addressed TSC should not be viewed, however, as a sort of information integration (fusion) approach. As is known, the fusion approach is a technique for merging (fusing, melting) information obtained from heterogeneous sources with different conceptual representations [54–56]. The TSC contains information from two heterogeneous sources, statistical and physics-of-failure based, and is not aimed at merging information. Each step in the TSC has a different objective and different time of application.

4.11.3 TSC in Modeling Aerospace Electronics Reliability

Technical diagnostics should be carried out as the *first step* of the TSC, with an objective to identify the faulty (malfunctioning) object or objects from the detected signals, "symptoms of faults (SoFs)." This can be done, particularly, using the Bayes formula. This powerful statistical means has been widely used for many years in numerous applications to update beliefs in many areas of engineering and applied science. At the *second step* the BAZ model is suggested to be employed to assess the RUL of the malfunctioning device. This model proceeds from the rationale that although the process of damage accumulation is highly temperature dependent, this process is affected primarily by various external loadings (stressors, stimuli) of the relevant nature, such as, for example, mechanical or thermal stress, voltage, current, light input, ionizing radiation, and so on. The model is based on the recognition of the experimentally observed situation that the breakage of the chemical bonds in a material under stress is due primarily to an external loading. Temperature still plays an important role though, mostly as the degradation (aging) factor. The *third step* of the TSC is aimed at updating reliability from the observed (but not anticipated) failures. It is another statistical means. Four parametric beta-distribution is suggested to be used to update reliability of the faulty object when/if it fails at the end of the anticipated or actual RUL [30,31]. The above three steps are addressed in some detail in the following text.

4.11.3.1 Step 1: Bayes Formula as a Suitable Technical Diagnostics Tool

The technical diagnostics effort is the TSC's first step. The objective of technical diagnostics is to recognize, in a continuous fashion and by using nondestructive means, the object's technical state and to assess the object's ability to continue its performance in the expected (specified) fashion. Technical diagnostics establishes the links between the observed (detected) signals, SoFs, and the underlying hidden state ("health") of the object. Technical diagnostics is focused on the most vulnerable elements (weakest links) of the design and can make use of the FOAT data collected earlier, at the DfR stage. Technical diagnostics encompasses a broad spectrum of problems associated with obtaining, processing, and assessing diagnostics information, including diagnostics models, decision-making rules, and algorithms. Technical diagnostics provides information for the subsequent PHM effort. Technical diagnostics is supposed to come up with the appropriate solutions and recommendations in the conditions of uncertainty and typically on the basis of limited information. The methods, techniques, and algorithms of technical diagnostics are based on probabilistic risk analysis (i.e., are used to quantify, on a probabilistic basis, the obtained information. Technical diagnostics provides assistance in making a decision as to whether the device/devices of interest is/are still sound or has/have become acceptably or unacceptably

Probabilistic Aerospace Electronics Reliability Engineering 93

faulty. There is always a risk that the interpretation of the obtained SoF signal might be a false alarm or, on the contrary, might lead to a "missing-the-target" decision.

4.11.3.2 Step 2: BAZ Equation as Suitable Physics-of-Failure Tool

The BAZ model is a generalization of the 1965 Zhurkov's [40] solid-state physics model, which is a generalization of the 1889 Arrhenius' [44] chemical kinetics model, which is, in its turn, a generalization of the 1886 Boltzmann's ("Boltzmann statistics") [43] model in the kinetic theory of gases. According to Boltzmann, the measurement of the heat contained in a gas is merely a measure of the motions of the individual particles that make up the gas. Three years later Arrhenius, member of the Boltzmann's team, determined that Boltzmann's model could be used to describe the temperature dependence not only of physical processes in a gas, but of chemical reactions as well. Arrhenius argued that reactants must first acquire a minimum amount of energy (it was he who introduced the term *activation energy*) U_0 to get transformed into chemical products. Arrhenius suggested that the percentage of molecules that had kinetic energy greater than U_0 at an absolute temperature T was calculated from the Boltzmann's distributions in statistical physics. Arrhenius' concept and reasoning have been widely accepted and widely applied later on not only to chemical reactions, but also to various physical processes involving solids. The inventors of the transistor, Bardeen, Brattain, and Shockley and the inventors of the integrated circuit, Kilby and Noyce, were all physicists, and this is perhaps one of the main reasons why Arrhenius' model has been widely used during the last 50 years or so in various reliability physics-related problems to assess the MTTF of semiconductor materials and devices.

Zhurkov [40] extended Arrhenius' model for the situation, when a solid body experienced the combined action of two major stimuli—elevated temperature and external tensile mechanical load. He proceeded from the fact that failure (fracture) of a solid was due primarily to an external mechanical tensile load that reduced the effective activation energy. The fracture toughness of the material was enhanced by material weakening (aging) that should be primarily attributed to an elevated temperature. Zhurkov and his associates tested over a long period of time a huge number of specimens made of about 50 different materials: metals, alloys, nonmetallic crystals, and even polymers. They came to the conclusion that the model (4.20) held well for all the materials tested and that the linear relationship (4.21) for the effective activation energy determines the sharp acceleration of the flaw propagation process in a body under stress. The physically meaningful BAZ model (4.2) enables one to evaluate the lifetime τ from the known applied external stress σ, the absolute temperature T, the time constant τ_0, the (stress-independent) binding (activation) energy U_0, and the stress-sensitivity factor γ. When the stress σ is mechanical stress, the sensitivity factor γ depends on the level of disorientation of the molecular structure of the material, the product $\gamma\sigma$ is the stress per unit volume and is characterized therefore by the same units as the activation energy U_0. If the load σ is not a mechanical stress [40,41], the units of the factor γ should be such that the product $\gamma\sigma$ is still expressed in energy units.

The Arrhenius model has well-established roots in the physics of failure of semiconductor devices, and many reliability physics models use the Arrhenius equation as their core. While Bayes formula, the first step in the TSC, does not anticipate or require any information about the physical nature of the obtained diagnostics signals, the BAZ formula has a clearly expressed physics-of-failure nature: when there is a reason to believe that the combination of the elevated temperature and an elevated stress could have led to the detected malfunction of a material, device, or a system, the additional information about the possible source of the deviation from the normal operation conditions and the assessment of the RUL of the product can be obtained and quantified using the BAZ model. When the lifetime τ in the BAZ model (4.2) is viewed as the MTTF, and the exponential law of reliability

$$P = \exp(-\lambda t) = \exp\left[-\frac{t}{\tau_0}\exp\left(-\frac{U_0 - \gamma\sigma}{kT}\right)\right] \qquad (4.35)$$

is used to evaluate the probability of nonfailure during the steady-state operation of the system, when possible failures are rare and random, the MTTF corresponds to the moment of time when the entropy of the law (4.20) reaches its maximum value [40,51]. In the formula (4.35), λ is the failure rate during the steady-state operation of the device. Indeed, the physical meaning of the distribution (4.35) can be seen from the formulas

$$\frac{\partial P}{\partial \lambda} = -\frac{H(P)}{\lambda}, \quad \frac{\partial P}{\partial t} = -\frac{H(P)}{t}, \quad \frac{\partial P}{\partial U_0} = \frac{H(P)}{kT}, \quad \frac{\partial P}{\partial \sigma} = -\frac{H(P)}{kT}\gamma = -\gamma\frac{\partial P}{\partial U_0}, \quad (4.36)$$

which follow from (4.35). In these formulas, $H(P) = -P\ln P$ is the entropy of the probability (4.35) of nonfailure. The formula (4.36) indicate particularly that the decrease in the probability of nonfailure with an increase in the failure rate is proportional to the level of the entropy of this probability and is inversely proportional to the level of the failure rate; that the decrease in the probability of nonfailure with time is proportional to the entropy of this probability and inversely proportional to the time in operation; that the decrease in the probability of nonfailure with an increase in the stress-free activation energy is proportional to the entropy of this probability and is inversely proportional to the absolute temperature; and that the stress sensitivity factor can be found as the ratio of the derivative of the probability of nonfailure with respect to the applied stress to the derivative of this probability with respect to the stress-free activation energy.

Probabilistic Aerospace Electronics Reliability Engineering

Equation (4.35) contains three empirical parameters: the stress-free activation energy, U_0; the external stress sensitivity factor, γ; and the time constant, τ_0. These parameters could be obtained on a stage-by-stage basis from the conducted FOAT. At the first stage, one should run the FOAT at two temperature levels, T_1 and T_2, while keeping the same level of the external stress, σ. The times t_1 and t_2 of failure should be recorded, when fractions $Q_1 = 1 - P_1$ and $Q_2 = 1 - P_2$ of the tested device population fail. The failure could be defined as a structural failure (crack, delamination, excessive deformation, etc.) or as a functional (electrical, optical, thermal) failure, when, for example, the detected leakage current [57,58] exceeds a certain allowable level, or when another suitable and physically meaningful and trustworthy criterion of failure (degradation) is applied. Then the following transcendental equation for the time ratio $\tau_1 = \dfrac{\tau_0}{t_1}$ can be obtained from Equation (4.35):

$$f(\tau_1) = \frac{\ln\left[-\tau_1 \dfrac{t_1}{t_2} \ln(1 - Q_2)\right]}{\ln\left[-\tau_1 \ln(1 - Q_1)\right]} - \frac{T_1}{T_2} - 0. \tag{4.37}$$

The solution to this equation can be obtained either by trial and error, assuming different τ_1 values and interpolating between those that are close to and slightly above and below zero, or, for higher accuracy, by using Newton's iterative process for solving transcendental equations. In the case in question, Newton's formula results in the recurrent relationship

$$\tau_{n+1} = \tau_n \left[1 + \frac{\ln(\beta\tau_n)}{1 - \dfrac{1-c}{f(\tau_n)}}\right], \tag{4.38}$$

which can be used to solve the following transcendental equations of the type (4.6) (subscript "1" is omitted):

$$f(\tau) = \frac{\ln(\alpha\tau)}{\ln(\beta\tau)} - c = 0. \tag{4.39}$$

After the time ratio $\tau = \tau_1 = \dfrac{\tau_0}{t_1}$ is found, then other parameters of interest can be determined with no difficulties.

Using a simple example, let us show that for the case when the objects of interest are subjected to an elevated temperature and, in addition, are loaded by two stressors: elevated humidity, H, and high voltage V. Let us assume that the failure rate, which characterizes the propensity of the material or the device to failure, is determined by the level of the leakage current: $\lambda = \gamma_1 I$.

Leakage current (current that flows from a conductive portion of a device to a portion that is intended to be nonconductive under normal conditions) is often used, as is known, as a possible degradation (failure) indicator [57,58]. Using the type (4.36) model, one could seek the probability of nonfailure of a product in the form:

$$P = \exp\left[-\gamma_I It \exp\left(-\frac{U_0 - \gamma_H H - \gamma_V V}{kT}\right)\right],\tag{4.40}$$

where the γ_H and γ_V factors reflect the sensitivities of the product to the corresponding stimuli (stressors). Although only two stressors were selected here, the model can be easily made multiparametric (i.e., generalized for as many stressors as necessary). The sensitivity factors γ are to be determined from the FOAT when the combined effect of all the stressors of importance is considered. To determine the sensitivity factors in the formula (4.41), one should run first the FOAT for two different temperatures, T_1 and T_2, while keeping the levels of the humidity, H, and the voltage, V, the same. Let the recorded percentages (fractions) of the failed samples be Q_1 and Q_2, respectively. Assuming a certain criterion of failure (say, when the level of the measured leakage current exceeds a certain level, I_*) and recording the corresponding times t_1 and t_2 to failure, one could obtain the following two relationships for the probability of failure:

$$Q_1 = 1 - \exp\left[-\gamma_I I_* t_1 \exp\left(-\frac{U_0 - \gamma_H H - \gamma_V V}{kT_1}\right)\right],$$

$$Q_2 = 1 - \exp\left[-\gamma_I I_* t_2 \exp\left(-\frac{U_0 - \gamma_H H - \gamma_V V}{kT_2}\right)\right].\tag{4.41}$$

Since the numerators, $U_0 - \gamma_H H - \gamma_V V$, in these formulas are kept the same, the sought sensitivity factor γ_I of the leakage current can be found from the transcendental equation:

$$f(\gamma_I) = \ln\left(-\frac{\ln(1 - Q_1)}{I_* t_1 \gamma_I}\right) - \frac{T_2}{T_1}\ln\left(-\frac{\ln(1 - Q_2)}{I_* t_2 \gamma_I}\right) = 0.\tag{4.42}$$

More than two series of FOAT tests and at more than two temperature levels should be conducted, so that the sensitivity parameter, γ_I, is validated with sufficient accuracy. This should be done to make sure that the relationship (4.40) works satisfactorily also beyond the region at which the FOAT is conducted. At the next stage, FOAT at two humidity levels, H_1 and H_2, should be carried out for the same temperature and voltage levels. This leads to the following equation:

Probabilistic Aerospace Electronics Reliability Engineering

$$\gamma_H = \frac{kT}{H_1 - H_2}\left[\ln\left(\frac{\ln(1-Q_1)}{I_* t_1 \gamma_I}\right) - \ln\left(\frac{\ln(1-Q_2)}{I_* t_2 \gamma_I}\right)\right]. \tag{4.43}$$

Similarly, by changing the voltages, V_1 and V_2, in the next set of FOAT one could find:

$$\gamma_V = \frac{kT}{V_1 - V_2}\left[\ln\left(-\frac{\ln(1-Q_1)}{I_* t_1 \gamma_I}\right) - \ln\left(-\frac{\ln(1-Q_2)}{I_* t_2 \gamma_I}\right)\right]. \tag{4.44}$$

Finally, the stress-free activation energy can be computed as

$$U_0 = \gamma_H H + \gamma_V V - kT \ln\left(-\frac{\ln(1-Q)}{I_* t \gamma_I}\right). \tag{4.45}$$

Example 4.3. Let, for example, the following input information be available: (1) After $t_1 = 35\,\text{h}$ of testing at the temperature $T_1 = 60°\text{C} = 333°\text{K}$, the voltage $V = 600\,\text{V}$ and the relative humidity $H = 0.85\%$, 10% of the tested modules exceeded the allowable (critical) level of the leakage current of $I_* = 3.5\,\mu\text{A}$ and, hence, failed, so that the probability of nonfailure is $P_1 = 0.9$. (2) After $t_2 = 70\,\text{h}$ of testing at the temperature $T_2 = 85°\text{C} = 358°\text{K}$ at the same voltage and the same relative humidity, 20% of the tested samples reached or exceeded the critical level of the leakage current and, hence, failed, so that the probability of nonfailure is $P_2 = 0.8$. Then Equation (4.42) results in the following transcendental equation for the leakage current sensitivity factor γ_I:

$$f(\gamma_I) = \ln\left(\frac{0.10536}{\gamma_I}\right) - 1.075075\ln\left(-\frac{0.22314}{\gamma_I}\right) = 0.$$

This equation yields $\gamma_I = 4926\,\text{h}^{-1}(\mu\text{A})^{-1}$. Thus, $\gamma_I I_* = 17,241\,\text{h}^{-1}$. A more accurate solution can always be obtained by using the Newton iterative method for solving transcendental equations. This concludes the *first step of testing*.

At the *second step*, tests at two relative humidity levels H_1 and H_2 were conducted for the same temperature and voltage levels. This led to the following relationship:

$$\gamma_H = \frac{kT}{H_1 - H_2}\left[\ln\left(-0.5800\times 10^{-4}\frac{\ln P_1}{t_1}\right) - \ln\left(-0.5800\times 10^{-4}\frac{\ln P_2}{t_2}\right)\right].$$

For example, after $t_1 = 40\,\text{h}$ of testing at the relative humidity of $H_1 = 0.5$ at the given voltage (say, $V = 600\,\text{V}$) and temperature (say, $T = 60°\text{C} = 333°\text{K}$), 5% of the tested modules failed, so that $P_1 = 0.95$, and after $t_2 = 55\,\text{h}$ of testing at the same temperature and at the relative humidity of $H = 0.85\%$, 10% of the tested modules failed, so that $P_2 = 0.9$. Then the above equation for the γ_H value, with the Boltzmann constant $k = 8.61733 \times 10^{-5}$ eV/K, yields $\gamma_H = 0.03292$ eV.

At the *third step*, FOAT at two different voltage levels $V_1 = 600$ V and $V_2 = 1000$ V have been carried out for the same temperature-radiation bias, say, $T = 85°C = 358°K$ and $H = 0.85$, and it has been determined that 10% of the tested devices failed after $t_1 = 40$ h of testing ($P_1 = 0.9$) and 20% of devices failed after $t_2 = 80$ h of testing ($P_2 = 0.8$). The voltage sensitivity factor can be found then as follows:

$$\gamma_V = \frac{0.02870}{400}\left[\ln\left(-0.5800 \times 10^{-4}\frac{\ln P_2}{t_2}\right) - \ln\left(-0.5800 \times 10^{-4}\frac{\ln P_1}{t_1}\right)\right]$$

$$= 4.1107 \times 10^{-6}\, \text{eV/V}.$$

After the sensitivity factors of the leakage current, the humidity, and the voltage are found, the stress-free activation energy can be determined for the given temperature and for any combination of loadings (stimuli). The third term in the equation for the stress-free activation energy plays the dominant role, so that, in approximate evaluations, only this term could be considered. Calculations indicate that the loading free activation energy in the above numerical example (even with the rather tentative, but still realistic, input data) is about $U_0 = 0.4770$ eV. This result is consistent with the existing experimental data. Indeed, for semiconductor device failure mechanisms, the activation energy ranges from 0.3 to 0.6 eV, for metallization defects and electro-migration in Al it is about 0.5 eV, for charge loss it is on the order of 0.6 eV, for Si junction defects it is 0.8 eV.

4.11.3.3 Step 3: Beta-Distribution as a Suitable Reliability Update Tool

The estimated probability, P, of nonfailure can be treated itself as a random variable, and its probabilistic characteristics can be determined from the appropriate probability distribution. Beta-distribution is widely used for this purpose. The "successes" and "failures" in beta-distribution could be any physically meaningful (indicative) criteria. It could be, for example, the number of samples (including redundancies, whose probabilities of nonfailure are never 100%) that survived the FOATs or QTs or, in the case of active redundancies, those that exhibited failures during FOAT or in actual operation. It could be also the allowable levels of temperature, or the power of the dynamic response of the device to the input random vibration power spectrum, or the level of the drop in the light output of a laser, or the level of leakage current. Such a generalization enables one to use the predictions based on the application of the beta-distribution, which is a powerful and flexible means for updating reliability.

The formal justification for using beta-distribution to update reliability information is based on the Bayesian theory, on one hand, and on the notion of conjugate distributions in this theory, on the other. If the posterior distribution of

Probabilistic Aerospace Electronics Reliability Engineering 99

the probability of nonfailure remains in the same family as its prior distribution, then the prior and the posterior distributions are called conjugate distributions. For certain choices of the prior (conjugate prior), the posterior distribution has the same algebraic form as the prior, although generally with different parameter values. There are many distributions that are conjugate ones. When it comes to updating reliability, the two-parametric beta-distribution defined on the interval [0,1] is considered the most easy to use.

> **Example 4.4.** The objective of the numerical example that follows is to illustrate how the suggested TSC can be used to assess and to maintain high probability of nonfailure in actual operation conditions.

4.11.3.4 Step 1: Application of Bayes Formula

The application of the Bayes formula enables one to assess the reliability of a particular malfunctioning device from the available general information for similar devices. See Example 2.20.

4.11.3.5 Step 2: Application of BAZ Equation

Assume that FOAT has been conducted at the design stage with an objective of determining the process parameters anticipated by the BAZ model, and that the first stage tests have been carried out at two temperature levels, T_1 and T_2, with the temperature ratio of $\dfrac{T_1}{T_2} = 0.95$ and the recorded time-to-failure ratio $\dfrac{t_1}{t_2} = 1.5$, until, say, half of the population of the devices failed: $Q_1 = Q_2 = 0.5$. Then Equation (4.5) results in the following equation for the sought dimensionless time $\tau_1 = \dfrac{\tau_0}{t_1}$:

$$f(\tau_1) = \frac{\ln(1.0397\tau_1)}{\ln(0.6931\tau_1)} - 0.95 = 0.$$

This equation has the following solution: $\tau_1 = \dfrac{\tau_0}{t_1} = 4.2250 \times 10^{-4}$, obtained by the trial-and-error (interpolation) technique. If Newton's formula is used, by putting, for example, $\tau_0 = 10^{-4}$ as the initial (zero) approximation and using the recurrent formula (4.6) to compute higher approximations, we obtain

$$\tau_1 = 2.73194 \times 10^{-4}; \quad \tau_2 = 4.06515 \times 10^{-4}; \quad \tau_3 = 4.32841 \times 10^{-4}; \quad \tau_4 = 4.33475 \times 10^{-4}.$$

The latter result agrees with the result obtained using the trial-and-error technique.

Let the FOAT be conducted at the temperature of $T = 450°K$ at two stress levels with the stress ratio of, say, $\dfrac{\sigma_2}{\sigma_1} = 1.2$. Testing is run until half of the population failed ($Q_1 = Q_2 = 0.5$), and the recorded time ratio, when failures occurred, has been $\dfrac{t_1}{t_2} = 1.5$. In this example, it is assumed that the time constant τ_0 in the BAZ equation is known from the previous FOAT. With this constant known, we calculate the $\dfrac{\tau_0}{t_1}$ ratio for the new time t_1. Let this ratio be $\dfrac{\tau_0}{t_1} = 4.0 \times 10^{-4}$. Then the loading σ_1 related energy is

$$\gamma\sigma_1 = \frac{6.21315 \times 10^{-21}}{-0.2} \left[\ln\left(4.0 \times 10^{-4} \times 0.6931\right) - \ln\left(4 \times 10^{-4} \times 1.5 \times 0.6931\right) \right]$$

$$= 0.125962 \times 10^{-19} \text{ J} = 0.07862 \text{ eV}$$

and the temperature T related energy kT is

$$kT = 1.3807 \times 10^{-23} \text{ J/°K} \times 450°K = 6.21315 \times 10^{-21} \text{ J}.$$

The ratio of the loading-related energy to the temperature-related energy is, therefore,

$$\frac{\gamma\sigma_1}{kT} = \frac{0.125962 \times 10^{-19}}{6.21315 \times 10^{-21}} = 2.02734.$$

This ratio will be larger for larger loadings and lower temperatures.

The ratio of the stress-free activation energy to the thermal energy can be determined as

$$\frac{U_0}{kT} = \frac{\gamma\sigma_1}{kT} - \ln\left[-\frac{\tau_0}{t_1} \ln\left(1 - Q\right) \right] = 2.02734 - \ln\left(-4 \times 10^{-4} \ln 0.5\right) = 10.2179.$$

Hence, the stress-free activation energy is

$$U_0 = \frac{U_0}{kT} kT = 10.2179 \times 6.21315 \times 10^{-21} \text{ J} = 0.634853 \times 10^{-19} \text{ J} = 0.3962 \text{ eV}.$$

The effective activation energy, when the stress, σ_1, is applied, is

$$U = U_0 - \gamma\sigma_1 = 0.3962 - 0.0786 = 0.3176 \text{ eV}.$$

The effective activation energy, when the stress, σ_2, is applied, is

$$U = U_0 - \gamma\sigma_2 = U_0 - \gamma\sigma_1 \frac{\sigma_2}{\sigma_1} = 0.3962 - 0.07862 \times 1.2 = 0.3019 \text{ eV}.$$

Let us assume that the FOAT-based and BAZ-based calculations carried out at the operation temperature of $T = 90°C = 363°K$ have indicated that the time factor is $\tau_0 = 10^{-4}$ s; the ratio of the stress-free activation energy to the temperature-related energy is $\frac{U_0}{kT} = 30.0$; and the ratio of the stress-related energy to the thermal energy is $\frac{\gamma\sigma}{kT} = 1.0$. Then the BAZ formula (4.20) results in the following projected lifetime:

$$\tau = \tau_0 \exp\left(\frac{U_0 - \gamma\sigma}{kT}\right) = 10^{-4} \exp(30.0 - 1.0) = 12.4662 \text{ years}.$$

This time decreases to

$$\tau = \tau_0 \exp\left(\frac{U_0 - \gamma\sigma}{kT}\right) = 10^{-4} \exp(30.0 - 1.2) - 0.8162 \times 10^6 \text{ s} = 10.2064 \text{ years}$$

for the 20% increase in the power of the vibration spectrum and is only

$$\tau = \tau_0 \exp\left(\frac{U_0 - \gamma\sigma}{kT}\right) = 10^4 \exp\left(\frac{29.0}{1.2}\right) = 36.2 \text{ days}$$

in the case of the 20°C increase in temperature. Thus, the increase in temperature should be in this example of a greater concern than the increase in the vibration response (in the output vibration spectrum). Also, based on the Bayes formula prediction, the malfunction of the device due to the increased temperature is more likely than because of the faulty vibration protection system.

As another example of the application of the multiparametric BAZ equation, let us consider the following situation.

After $t_1 = 35$ h of testing at temperature $T_1 = 60°C = 333°K$ at the given voltage $V = 600$ V and the given relative humidity level $H = 85\% \ RH = 0.85$, 10% of the tested samples have reached or exceeded the critical level of the leakage current of $I_* = 3.5 \ \mu A$ and, hence, have failed; thus, the probability of nonfailure is $P_1 = 0.9$. In addition, the other set of FOAT has indicated that, after $t_2 = 70$ h of testing at a temperature of $T_2 = 85°C = 358°K$ at the same voltage and the same humidity level, 20% of the tested samples reached or exceeded the critical level of the leakage current and because of that failed, so that the probability of nonfailure is $P_2 = 0.8$. With these input data, the sensitivity factor for the leakage current can be found from the equation

$$f(\gamma_I) = \ln\left(\frac{0.10536}{\gamma_I}\right) - 1.075075 \times \ln\left(-\frac{0.22314}{\gamma_I}\right) = 0$$

and is $\gamma_I = 4926\,\mathrm{h}^{-1}(\mu\mathrm{A})^{-1}$. The product of the leakage current factor and the acceptable level of this current is, therefore, $\gamma_I I_* = 1721\,\mathrm{h}^{-1}$. At the second stage, FOAT at two humidity levels, H_1 and H_2, has been conducted while keeping the same temperature and voltage levels. This leads to the relationship:

$$\gamma_H = \frac{kT}{H_1 - H_2}\left[\ln\left(-0.5800 \times 10^{-4}\,\frac{\ln P_1}{t_1}\right) - \ln\left(-0.5800 \times 10^{-4}\,\frac{\ln P_2}{t_2}\right)\right].$$

Let $t_1 = 40\,\mathrm{h}$ of testing at the relative humidity level of $H_1 = 50\%RH = 0.5$ at the given voltage of $V = 600\,\mathrm{V}$ and temperature $T = 60°C = 333°K$, 5% of the tested samples have failed so that $P_1 = 0.95$. The FOAT carried out for the second set of samples has indicated that after $t_2 = 55\,\mathrm{h}$ of testing at the same temperature and at the same relative humidity level of $H_1 = 85\%RH = 0.85$, 10% of the tested samples failed, so that the probability of nonfailure is $P_2 = 0.9$. Then, the above equation, with $k = 8.61733 \times 10^{-5}\,\mathrm{eV/K}$, results in the following value of the sensitivity factor for the relative humidity: $\gamma_H = 0.03292\,\mathrm{eV}$. At the third stage, FOAT at two different voltage levels, $V_1 = 600\,\mathrm{V}$ and $V_2 = 1000\,\mathrm{V}$, was carried out for the same temperature-humidity bias of $T = 85°C = 358°K$ and $H = 85\%RH = 0.85$, as in the previous example. It has been determined that 10% of the tested samples failed after $t_1 = 40\,\mathrm{h}$ of testing ($P_1 = 0.9$), and 20% of devices failed after $t_2 = 80\,\mathrm{h}$ of testing ($P_2 = 0.8$). The voltage sensitivity factor can be then found as

$$\gamma_V = \frac{0.02870}{400}\left[\ln\left(-0.5800 \times 10^{-4}\,\frac{\ln P_2}{t_2}\right) - \ln\left(-0.5800 \times 10^{-4}\,\frac{\ln P_1}{t_1}\right)\right]$$

$$= 4.1107 \times 10^{-6}\,\mathrm{eV/V},$$

After the factors of the leakage current, the relative humidity, and the voltage have been determined, the stress-free activation energy, U_0, can be computed for the given temperature and for any combination of loadings on the basis of formula (4.20). Calculations indicate that the loading-free activation energy in this numerical example is about $U_0 = 0.4770\,\mathrm{eV}$.

Thus, the output of this TSC stage is the assessed, on the probabilistic basis, the RUL of the device(s). As has been indicated in the abstract, if the assessed RUL time is still long enough, no action might be needed, if not, corrective restoration action becomes necessary. In any event, after the first two TSC steps have been carried out, the devices are put back into operation, provided that the assessed probability of their continuing failure-free operation is found to be satisfactory. If failure nonetheless occurs, the third step should be undertaken to update the predicted reliability. Statistical beta-distribution, in which the probability of failure is treated as a random variable, is suggested to be used at the third step.

Probabilistic Aerospace Electronics Reliability Engineering

4.11.3.6 Step 3: Application of Beta-Distribution

See Example 2.28.

4.12 Conclusions

The following conclusions can be drawn from the carried out analysis:

- A combined statistics-based and physics-of-failure-based TSC is suggested as a possible methodology for the assessment of the aerospace electronics reliability at different stages of the product's life: at the design stage, at the accelerated testing stage, at the fabrication stage, and certainly at the operation stage. The concept enables one to obtain, on a continuous and cost-effective basis, the most trustworthy and comprehensive information about product's reliability.

- The concept can also be employed when an attempt to optimize reliability is undertaken (i.e., when trying to find the most feasible compromise between reliability, time to market, and cost-effectiveness). It goes without saying that not all the three steps of the TSC addressed should be used in each particular reliability-related effort.

- The application of FOAT, the PDfR concept, and particularly the multiparametric BAZ model enables one to improve dramatically the state of the art in the field of the electronic products reliability prediction and assurance.

- Since FOAT cannot do without simple, easy-to-use, and physically meaningful predictive modeling, the role of such modeling, both computer aided and analytical (mathematical), is making the suggested new approach to QT practical and successful.

- It is imperative that the reliability physics that underlies the mechanisms and modes of failure is well understood. Such an understanding can be achieved only provided that flexible, powerful, and effective PDfR efforts are implemented.

References

1. J. Sloan, *Design and Packaging of Electronic Equipment*, Van Nostrand Reinhold Company, New York, 1985.
2. R. R. Tummala, *Microelectronics Packaging Handbook*, Springer, New York, 1988.

3. MIL-HDBK-338B, October 1, 1998.
4. D. Crowe and A. Feinberg (eds.), *Design for Reliability* (Electronic Handbook Series), Springer, New York, 2001.
5. E. Suhir, Y.C. Lee, and C.P. Wong (eds.), *Micro- and Opto-Electronic Materials and Structures: Physics, Mechanics, Design, Reliability, Packaging*, vols 1 and 2, Springer, New York, 2007.
6. A. Bensoussan and E. Suhir, "Design-for-Reliability of Aerospace Electronics: Attributes and Challenges." In *IEEE Aerospace Conference*, Big Sky, Montana, March 2013.
7. E. Suhir, "Probabilistic Design for Reliability," *ChipScale Reviews*, vol. 14, No. 6, 2010.
8. E. Suhir, R. Mahajan, A. Lucero, and L. Bechou, "Probabilistic Design for Reliability and a Novel Approach to Qualification Testing." In *IEEE Aerospace Conference*, Big Sky, Montana, 2012.
9. E. Suhir, "Statistics- and Reliability-Physics-Related Failure Processes," *Modern Physics Letters B*, in press vol. 28, No. 13, May 30, 2014, pp. 1–10.
10. E. Suhir, *Applied Probability for Engineers and Scientists*, McGraw-Hill, New York, 1997.
11. H. Czilos (ed.), *Handbook of Technical Diagnostics*, Springer, New York, 2013.
12. M. Pecht, *Product Reliability, Maintainability, and Supportability Handbook*, CRC Press, New York, 1995.
13. S. Mishra, M. Pecht, T. Smith, I. McNee, and R. Harris, "Remaining Life Prediction of Electronic Products Using Life Consumption Monitoring Approach." In *European Microelectronics Packaging and Interconnection Symposium*, Cracow, June 16–18, 2002.
14. A. Landzberg (ed.), *Microelectronics Manufacturing Diagnostics Handbook* (Electrical Engineering), Springer, New York, 1993.
15. P. Lall, N. Islam, P. Choudhary, and J. Suhling, "Prognostication and Health Monitoring of Leaded and Lead Free Electronic and MEMS Packages in Harsh Environments." In *55th IEEE Electronic Components and Technology Conference*, Orlando, FL, June 1–3, 2005.
16. N. Vichare and M. Pecht, "Prognostics and Health Management of Electronics," *IEEE CPMT Transactions*, vol. 29, No.1, 2006, pp. 222–229.
17. L. Nasser and M. Curtin, "Electronics Reliability Prognosis through Material Modeling and Simulation." In *IEEE Aerospace Conference*, Big Sky, Montana, March 2006, pp. 1107–1122.
18. P. Lall, N. Islam, K. Rahim, J. Suhling, and S. Gale, "Prognostics and Health Management of Electronic Packaging," *IEEE CPMT Transactions*, vol. 29, No. 3, 2006.
19. P. Bonissone, K. Goebel, and N. Iyer, "Knowledge and Time: Selected Case Studies in Prognostics and Health Management (PHM)." In *IPMU 2006 Industrial Session*, Paris, July 2–7, 2006.
20. J. Gu, N. Vichare, T. Tracy, and M. Pecht, "Prognostics Implementation Methods for Electronics." In *53rd Annual Reliability and Maintainability Symposium (RAMS)*, Orlando, FL, 2007.
21. J. Gu, D. Barker, and M. Pecht, "Prognostics Implementation of Electronics Under Vibration Loading," *Microelectronics Reliability*, vol. 47, No. 12, 2007, pp. 317–323.

22. P. Bonissone and N. Iyer, "Knowledge and Time: A Framework for Soft Computing Applications in PHM." In D. Bouchon-Meunier, C. Marsala, M. Rifqi, and R.R. Yager (eds.), *Uncertainty and Intelligent Information Systems*, World Scientific, 2007.

23. P. Bonissone, "Soft Computing Applications in Prognostics and Health Management: A Time and Knowledge Framework with Selected Case Studies." In *AAAI Fall Symposium on Artificial Intelligence for Prognostics*, Arlington, VA, November 9–11, 2007.

24. M.G. Pecht, *Prognostics and Health Management of Electronics*, John Wiley, New York, 2008.

25. M. Daigle, A. Saxena, and K. Goebel, "An Efficient Deterministic Approach to Model-Based Prediction Uncertainty Estimation." In *Annual Conference of the PHM Society*, September 2012.

26. S. Sankararaman and K. Goebel, "Uncertainty Quantification in Remaining Useful Life of Aerospace Components Using State Space Models and Inverse FORM," Paper No. AIAA-2013-1537. In *Proceedings of 15th Non-Deterministic Approaches Conference co-located with the 54th AIAA/ASME/ASCE/AHS/ASC Structures, Structural Dynamics and Materials Conference*, Boston, MA, April 8–11, 2013.

27. S. Sankararaman, M. Daigle, and K. Goebel, "Analytical Algorithms to Quantify the Uncertainty in Remaining Useful Life Prediction." In *2013 IEEE Aerospace Conference*. Big Sky, MT, March 3–8, 2013.

28. S. Sankararaman and K. Goebel, "Remaining Useful Life Estimation in Prognosis: An Uncertainty Propagation Problem." In *Proceedings of the 2013 AIAA Infotech@Aerospace Conference, co-located with the AIAA Aerospace Sciences—Flight Sciences and Information Systems Event*, Boston, MA, August 19–22, 2013.

29. S. Sankararaman and K. Goebel, "Uncertainty in Prognostics: Computational Methods and Practical Challenges." In *2014 Aerospace Conference*, Big Sky, MT, March 1–8, 2014.

30. G. Casewell, "Using Physics of Failure to Predict System Level Reliability for Avionic Electronics." In *2014 Aerospace Conference*, Big Sky, MT, March 1–8, 2014.

31. P. Bonissone, "Soft Computing Applications in PHM." In *Proceedings of FLINS2008*, Madrid, Spain, in Computational Intelligence in Decision and Control, DaRua, Montero, Lu, Martinez, Dhondt, Kerre, eds., World Scientific, 2008.

32. T. Bayes and R. Price, "An Essay towards Solving a Problem in the Doctrine of Chance. By the late Rev. Mr. Bayes, communicated by Mr. Price, in a letter to John Canton, M. A. and F. R. S," *Philosophical Transactions of the Royal Society of London*, vol. 53, 1763, pp. 1–11.

33. S.M. Stigler, "Who Discovered Bayes' Theorem?" *The American Statistician*, vol. 37, No. 4, 1983, pp. 290–296.

34. E.T. Jaynes, "Bayesian Methods: General Background." In J.H. Justice (ed.), *Maximum-Entropy and Bayesian Methods in Applied Statistics*, Cambridge University Press, Cambridge, 1986.

35. A.W.F. Edwards, "Is the Reference in Hartley (1749) to Bayesian Inference?" *The American Statistician*, vol. 40, No. 2, 1986, pp. 109–110.

36. L. Daston, *Classical Probability in the Enlightenment*, Princeton University Press, Princeton, NJ, 1988.
37. J.V. Bernardo and A.F.M. Smith, *Bayesian Theory*, Wiley, New York, 1994.
38. S.E. Fienberg, *When Did Bayesian Inference Become "Bayesian"?*, 2006, pp. 1–40.
39. S.B. McGrayne, *The Theory That Would Not Die: How Bayes' Rule Cracked the Enigma Code, Hunted Down Russian Submarines, and Emerged Triumphant from Two Centuries of Controversy*, Yale University Press, New Haven, CT, 2011.
40. S.N. Zhurkov, "Kinetic Concept of the Strength of Solids," *International Journal of Fracture Mechanics*, vol. 1, No. 4, 1965, pp. 247–249.
41. E. Suhir, "Boltzmann-Arrhenius-Zhurkov (BAZ) Model in Physics-of-Materials Problems," *Modern Physics Letters B*, vol. 27, 2013, pp. 1–15.
42. E. Suhir, "Predicted Reliability of Aerospace Electronics: Application of Two Advanced Probabilistic Techniques." In *IEEE Aerospace Conference*, Big Sky, MT, March 2013.
43. L. Boltzmann, "The Second Law of Thermodynamics. Populare Schriften,"Essay 3, Address to a Formal Meeting of the Imperial Academy of Science, May 29, 1886.
44. S. Arrhenius, "Ueber den Einfluss des Atmosphärischen Kohlensäurengehalts auf die Temperatur der Erdoberfläche." In *Proceedings of the Royal Swedish Academy of Science*, vol. 22, No. 1, 1896, pp. 1–101.
45. M. Abramowitz and I.A. Stegun, *Handbook of Mathematical Functions with Formulas, Graphs, and Mathematical Tables*, Dover, 1966.
46. R.J. Beckman and G.L. Tietjen, "Maximum Likelihood Estimation for the Beta Distribution," *Journal of Statistical Computation and Simulation*, vol. 7, Nos 3–4, 1978 pp. 253–258.
47. A.K. Gupta (ed.), *Handbook of Beta Distribution and Its Applications*, CRC Press, Boca Raton, FL, 2004.
48. E. Suhir and B. Poborets, "Solder Glass Attachment in Cerdip/Cerquad Packages: Thermally Induced Stresses and Mechanical Reliability," *ASME Journal of Electronic Packaging (JEP)*, vol. 112, No. 2, 1990, pp. 1–6.
49. E. Suhir, "Probabilistic Approach to Evaluate Improvements in the Reliability of Chip-Substrate (Chip-Card) Assembly," *IEEE CPMT Transactions, Part A*, vol. 20, No. 1, 1997, pp. 60–63.
50. E. Suhir, "Thermal Stress Modeling in Microelectronics and Photonics Packaging, and the Application of the Probabilistic Approach: Review and Extension," *IMAPS International Journal of Microcircuits and Electronic Packaging*, vol. 23, No. 2, 2000, pp. 261–267.
51. S. Radhakrishnan, G. Subbarayan, and L. Nguyen, "Probabilistic Physical Design of Fiber-Optic Structures." In E. Suhir, Y.C. Lee, and C.P. Wong (eds.), *Micro- and Opto-Electronic Materials and Structures: Physics, Mechanics, Design, Reliability and Packaging*, vol. 2, Springer, 2007, pp. 2706–2711.
52. E. Suhir, "Could Electronics Reliability Be Predicted, Quantified and Assured?" *Microelectronics Reliability*, vol. 53, No. 53, 2013, pp. 925–936.
53. E. Suhir, A. Bensoussan, J. Nicolics, and L. Bechou, "Highly Accelerated Life Testing (HALT), Failure Oriented Accelerated Testing (FOAT), and Their Role in Making a Viable Device into a Reliable Product." In *IEEE Aerospace Conference*, Big Sky, MT, March 2014.
54. H.B. Mitchell, *Multi-Sensor Data Fusion – An Introduction*, Springer-Verlag, Berlin, 2007.

55. S. Das, *High-Level Data Fusion*, Artech House Publishers, Norwood, MA, 2008.
56. E.P. Blasch, E. Bosse, and D.A. Lambert, *High-Level Information Fusion Management and System Design*, Artech House Publishers, Norwood, MA, 2012.
57. Leakage Current. Condor Application note 7/01,AN-113 (Condor DC Power Supplies, Inc. 2311, Stathan Parkway, Oxnard, CA 93033).
58. S. Amin, M. Amin, and M. Ali, "Monitoring of Leakage Current for Composite Insulators and Electrical Devices,"*Reviews on Advanced Materials Science*, vol. 21, 2009, pp. 75–89.

5

Probabilistic Assessment of an Aerospace Mission Outcome

5.1 Summary

A double-exponential probability distribution function (DEPDF) is intro-duced to quantify the likelihood of the human failure to perform his/her duties, when operating a vehicle: an aircraft, a spacecraft, a boat, a helicopter, a car, and so on. Such a failure, if any, is attributed in our approach to the insufficient human capacity factor (HCF), when there is a need to cope with a high (extraordinary, off-normal) level mental workload (MWL). A possible application of the suggested concept is a situation when an imperfect human, an imperfect equipment/instrumentation, and an uncertain and often harsh environment contribute jointly to the likelihood of a vehicular mission failure and/or insufficient safety. While the human's performance is characterized by the DEPDF, the performance of the equipment (instrumentation), which includes the performance of both the hardware and the software, is character-ized by the Weibull distribution, and the role of the uncertain environment is considered by the probability of the occurrence of environmental conditions of the anticipated level of severity. The suggested model and its possible modifi-cations and generalizations can be helpful, after appropriate sensitivity analy-ses are carried out, when developing guidelines for personnel selection and training; when choosing the appropriate simulation conditions; and/or when there is a need to decide, if the existing levels of automation and the employed equipment (instrumentation) are adequate in off-normal, but not impossible, situations. If not, additional and/or more advanced and perhaps more expen-sive equipment/instrumentation should be developed, tested, and installed.

5.2 Introduction

In the analysis that follows, we introduce a DEPDF [1–5] to quantify the like-lihood of a human nonfailure to perform his/her duties, when operating a vehicle (aircraft, spacecraft, ocean-going vessel, etc.). We consider a situation when an imperfect human, imperfect equipment, and an uncertain and often harsh environment contribute jointly to the outcome of a mission or to a likelihood of a mishap. We believe that the suggested concept and its generalizations, after the appropriate sensitivity analyses are carried out, can be helpful when developing guidelines for personnel selection and training; when choosing the appropriate flight simulation conditions; and/or when there is a need to decide if the existing level of automation and the existing navigation instrumentation and equipment are adequate in extraordinary (off-normal) situations. If not, additional and/or more advanced and perhaps more expensive instrumentation and equipment should be considered, developed, and installed.

Our analysis is an attempt to quantify, on the probabilistic basis, using analytical probabilistic risk analysis (PRA) techniques, the role that the human-in-the-loop plays, in terms of his/her ability (capacity) to cope with a mental overload. Using an analogy from the reliability engineering field and particularly with the "stress-strength" interference model (see, e.g., [1]), the MWL could be viewed as a certain "demand" ("stress"), while the HCF could be viewed as a "capacity" ("strength"). In our DEPDF model, we combine the demand and the capacity factors within the same probability of nonfailure distribution. The relative levels of the MWL and HCF determine in our concept the likelihood of a mission or an off-normal situation success and safety.

The MWL ("demand"/"stress") depends on the operational conditions and the complexity of the mission (i.e., has to do with the significance of the general task) [4–27]. The MWL is directly affected by the challenges that a navigator faces, when he/she has to control the vehicle in a complex, heterogeneous, multitask, and often uncertain and harsh environment. Such an environment includes numerous different and interrelated concepts of situational awareness: spatial awareness for instrument displays; system awareness (e.g., for keeping the pilot informed about actions that have been taken by automated systems); and task awareness that has to do with attention and task management.

As to the HCF ("capacity"/"strength"), it considers, but might not be limited to, professional experience and qualifications; capabilities and skills; level of training; performance sustainability; ability to concentrate; mature thinking; ability to operate effectively, in a "tireless" fashion, under pressure, and, if needed, for a long period of time (tolerance to stress); team-player attitude, when needed; swiftness in reaction, if necessary [3,4]; and so on. In the analysis pursued in this chapter, it is assumed that while the MWL and the HCF are random variables, the most likely ("specified") MWL and HCF values in a particular mission and for a particular individual are deterministic parameters that are known (established, predetermined) in advance. This could be done particularly by employing accelerated

Probabilistic Assessment of an Aerospace Mission Outcome 111

testing on flight simulator equipment. Testing should continue until an anticipated failure (whatever the definition) occurs and the mean time to failure (MTTF) is measured for the selected group of navigators. Such failure-oriented-accelerated testing (FOAT, "testing-to-fail"), as opposite to "testing-to-pass," known in reliability engineering as qualification testing [28–30], is viewed to be analogous to the accelerated life testing (ALT) in electronics and photonics engineering [28].

Although the evaluation of the most likely MWL and HCF is beyond the scope of the present analysis, a brief discussion is put, nonetheless, on how some factors affecting the specified MWL and HCF are or might be approached in today's practice. It is noteworthy that the ability to evaluate the "absolute" level of the MWL, important as it is for noncomparative evaluations, is less critical in our approach, aimed at the comparative assessments of the likelihood of a casualty in situations, when the relative levels of MWL and HCF are critical. We would like to point out also that we do not intend here to suggest any accurate, complete, ready-to-go, off-the-shelf-type methodology, but intend to show and to discuss how the PRA methods and approaches could be effectively employed to quantify the role of the human factor, when both human performance and equipment (instrumentation) behavior and reliability contribute jointly to the likelihood of an outcome of a mission or an extraordinary situation. We believe, however, that the taken approach is both new and general, and, with the appropriate modifications and generalizations, is applicable to many other situations, not even necessarily in the vehicular domain, when a human encounters an uncertain environment and/or a hazardous situation and/or interacts with never perfect hardware and software, and with an uncertain-and-harsh environment.

5.3 DEPDF of Human Nonfailure

We assume that the steady-state probability P^h (F,G) of the navigator's non-failure, when the vehicle is operated in off-normal (extraordinary) conditions, is distributed in accordance with the following double-exponential law of the EVD type [1–5]:

$$p^h(F,G) = P_0 \exp\left[\left(1 - \frac{G^2}{G_0^2}\right)\exp\left(1 - \frac{F^2}{F_0^2}\right)\right]. \tag{5.1}$$

Here P_0 is the probability of nonfailure of the human at the specified (normal) MWL level, when $G = G_0$; G_0 is the most likely (normal, specified) MWL (i.e.,

MWL in ordinary operation conditions; G is the actual (elevated, off-normal) MWL; $F = F_0$ is the most likely (normal, specified) HCF—that is, the HCF in ordinary (normal) conditions; and F is the actual HCF exhibited at the extraordinary (off-normal) conditions. The P_0 level of the probability of the human performance nonfailure in normal conditions—that is, in the case of a human with a normal (most likely) level of the HCF (a performer with ordinary skills in the profession)—should be established beforehand, as a function of the G_0 level when the HCF $F = F_0$). This could be done, for example, by conducting testing and measurements on a flight simulator. The calculated ratios

$$\bar{P} = \frac{P^h(F,G)}{P_0} = \exp\left[\left(1 - \frac{G^2}{G_0^2}\right)\exp\left(1 - \frac{F^2}{F_0^2}\right)\right] \tag{5.2}$$

of the probability of human nonfailure in off-normal conditions to the probability of nonfailure in normal conditions are shown in Table 5.1. The following conclusions that make physical sense can be drawn from the calculated data:

1. At normal MWL level ($G = G_0$) and/or at an extraordinarily (exceptionally) high HCF level the probability of human nonfailure is close to 100%.
2. The probabilities of human nonfailure in off-normal conditions are always lower than the probabilities of nonfailure in normal conditions. This obvious fact is quantified by the calculated data.
3. If the MWL is exceptionally high, the human will definitely fail, no matter how high his/her HCF is.

TABLE 5.1

Calculated $\bar{P} = P^h(F,G)/P_0$ Ratios of the Probability $P^h(F,G)$ of Human Nonfailure in Off-Normal Conditions to the Probability P_0 of Nonfailure in Normal Conditions

F^2/F_0^2 \ G^2/G_0^2	1	2	3	4	5	8	10	∞
1	1	0.3679	0.1353	0.0498	0.0183	9.1188E-4	1.234E-4	0
2	1	0.6922	0.4791	0.3317	0.2296	0.0761	0.0365	0
3	1	0.8734	0.7629	0.6663	0.5820	0.3878	0.2958	0
4	1	0.9514	0.9052	0.8613	0.8194	0.7057	0.6389	0
5	1	0.9819	0.9640	0.9465	0.9294	0.8797	0.8480	0
8	1	0.9991	0.9982	0.9978	0.9964	0.9936	0.9918	2.5E-40
10	1	0.9999	0.9998	0.9996	0.9995	0.9991	0.9989	4.4E-6
∞	1	1	1	1	1	1	1	1

Probabilistic Assessment of an Aerospace Mission Outcome 113

4. If the HCF is high, even a significant MWL has a small effect on the probability of nonfailure, unless this workload is exceptionally large.

5. The probability of nonfailure decreases with an increase in the MWL (especially for relatively low MWL levels) and increases with an increase in the HCF (especially for relatively low HCF levels). This intuitively obvious fact is quantified by the calculated data.

6. For high HCFs, the increase in the MWL level has a much smaller effect on the probabilities of nonfailure than for relatively low HCFs. Table 5.1 data show also that the increase in the F/F_0 ratio and in the G/G_0 ratio above the 3.0 value has a small effect on the probability of nonfailure. This means particularly that the navigator (pilot) does not have to be trained for an extraordinarily high MWL and does not have to be trained by a factor higher than 3.0 compared to a navigator of ordinary capacity (skills, qualification). In other words, a pilot does not have to be a superman/superwoman to successfully cope with a high-level MWL, but still has to be trained in such a way that, when there is a need, he/she would be able to cope with an elevated MWL that is by a factor of 3.0 higher than the normal level, and his/her HCF should be by a factor of 3.0 higher than what is expected of the same person or of a person with an ordinary HCF in ordinary (normal) conditions.

From (5.2) we find, by differentiation,

$$\frac{d\bar{P}}{dG} = -\frac{2H}{G}\frac{G^2}{G^2 - G_0^2},$$

(5.3)

where $H = -\bar{P}\ln\bar{P}$ is the entropy (see, e.g., [1]) of the distribution of the relative probability of the human nonfailure in extraordinary (off-normal) conditions of operation as compared to ordinary (normal) conditions. At the MWL levels close to the normal level, the change in the relative probability of nonfailure with the increase in the MWL level is significant. In another extreme case, when $G \ggg G_0$, we have

$$\frac{d\bar{P}}{dG} = -\frac{2H}{G}.$$

(5.4)

This formula explains the physical meaning of the DEPDF (4.1): the change in the probability of nonfailure with the change in the level of the MWL is proportional, for large enough MWL levels, to the uncertainty level (entropy of the distribution of this probability) and is inversely proportional to the

MWL level. The right part of the formula (5.4) could be viewed as a kind of a coefficient of variation (cov), where the role of the uncertainty level in the numerator is played by the entropy (rather than by the standard deviation), and the role of the stress (loading) level in the denominator is played by the MWL level (rather than by the mean value of the random characteristic of interest).

5.4 Likelihood of the Vehicular Mission Success and Safety

The success or failure of a vehicular mission could be time dependent and, in addition, could have different probabilities of success/failure at different stages (segments). Let, for example, the mission of interest consist of n consecutive segments ($i = 1,2,...,n$) that are characterized by different probabilities, q_i, of occurrence of a particular harsh environment or by other extraordinary conditions during the fulfillment of the ith segment of the mission; by different durations, T_i, of these segments; and by different failure rates, λ_i^e, of the equipment and instrumentation when navigation at these segments. These failure rates may or may not depend on the environmental conditions, but could be affected by the equipment/materials aging, degradation, and other time-dependent causes.

In the simplified example that follows, we assume that the combined input of the hardware and the software, as far as the failure rate of the equipment and instrumentation is concerned, is evaluated beforehand and is adequately reflected by the appropriate failure rate λ_i^e (failure rate of the equipment) values. These values could be either determined from the vendor specifications or obtained based on the specially designed and conducted accelerated testing and the subsequent predictive modeling [28]. The probability of the equipment nonfailure at the moment t_i of time during the flight (mission fulfillment) on the ith segment, assuming, in an approximate analysis, that Weibull distribution

$$P_i^e = \exp\left[-\left(\lambda_i^e t_i\right)^{\beta_i^e}\right] \tag{5.5}$$

is applicable. Here, $0 \le t_i \le T_i$ is an arbitrary moment of time during the fulfillment of the mission on the ith segment, and β_i^e is the shape parameter in the Weibull distribution. We assume that the time-dependent probability of the human performance nonfailure can be also represented in the form of Weibull distribution:

$$P_i^h(t_i) = P_i^h(0)\exp\left[-\left(\lambda_i^h t_i\right)^{\beta_i^h}\right], \tag{5.6}$$

Probabilistic Assessment of an Aerospace Mission Outcome 115

where λ_i^h is the failure rate, β_i^h is the shape parameter, and $P_i^h(0)$ is the probability of the human nonfailure at the initial moment of time $t_i = 0$ of the given segment. When $t_i \to \infty$, the probability of nonfailure (say, because of the human fatigue or other causes) tends to zero. The probability $P_i^h(0)$ of nonfailure at the initial moment of time can be assumed in the form (5.1):

$$P_i^h(0) = P_0 \exp\left[\left(1 - \frac{G_i^2}{G_0^2}\right)\exp\left(1 - \frac{F_i^2}{F_0^2}\right)\right]. \tag{5.7}$$

Then the probability of the mission failure at the ith segment can be found as

$$Q_i(t_i) = 1 - P_i^e(t_i) P_i^h(t_i). \tag{5.8}$$

Since

$$\sum_{i=1}^{n} q_i = 1 \tag{5.9}$$

(condition of normalization), the overall probability of the mission failure can be determined as

$$Q = \sum_{i=1}^{n} q_i Q_i(t_i) = 1 - \sum_{i=1}^{n} q_i P_i^e(t_i) P_i^h(t_i). \tag{5.10}$$

This formula can be used for the assessment of the probability of the overall mission failure, as well as, if necessary, for specifying the failure rates and the HCF in such a way that the probability of failure,when a human is involved, would be sufficiently low and acceptable. It can be used also, if possible, to choose an alternative route in such a way that the set of the probabilities q_i brings the overall probability of failure of the mission to the acceptable level. If at a particular segment of the fulfillment of the mission the human performance is not critical (e.g., during the take-off of a shuttle), then the corresponding probability $P_i^h(t_i)$ of human nonfailure should be put equal to one. But if there is confidence that the equipment (instrumentation) failure is not critical, or if there is a reason to believe that the probability of the equipment nonfailure is considerably higher than the probability of the human nonfailure, then it is the probability $P_i^e(t_i)$ of the equipment non-failure that should be put equal to one. Finally, if one is confident that a certain level of the harsh environment will be certainly/inevitably encountered during the fulfillment of the mission at the ith segment of the route, then the corresponding probability q_i should be put equal to one.

5.5 Equipment (Instrumentation) Failure Rate

Failure rate of the equipment (instrumentation) should be established, of course, based on the reliability physics of the particular underlying phenomenon. Examine, as suitable examples, two typical situations. If the possible failure of the vulnerable structural element of a particular piece of equipment, device, or a subsystem could be attributed to an elevated temperature and stress, then the Boltzmann–Arrhenius–Zhurkov (BAZ) law (addressed and discussed in detail in the previous chapter)

$$\tau = \tau_0 \exp\left(\frac{U - \gamma\sigma}{kT}\right) \tag{5.11}$$

can be used to assess the MTTF τ. In this formula, T is the absolute temperature, U is the activation energy, k is Boltzmann's constant, σ is the design stress (not necessarily mechanical) acting in the item of interest, and τ_0 and γ are empirical parameters that should be established (found) based on the specially designed and conducted ALTs. Actually, the activation energy U is also an empirical parameter, but, for various structural elements of silicon-based semiconductor electronic devices the activation energies have been determined and could be found in the literature. The second term in the numerator of the above formula accounts for the reduction in the activation energy level in the presence of a stress. This stress does not have to be a mechanical stress, but could be any stimulus or stimuli that affect the probability of failure of the instrumentation/equipment of importance. If stress is not considered, the above formula reduces to the well-known Arrhenius equation. After the MTTF τ is determined, the corresponding failure rate can be found as

$$\lambda = \frac{1}{\tau} = \frac{1}{\tau_0} \exp\left(-\frac{U - \gamma\sigma}{kT}\right) = \frac{Q_T}{\tau_0}, \tag{5.12}$$

where

$$Q_T = \exp\left(-\frac{U - \gamma\sigma}{kT}\right) \tag{5.13}$$

is the steady-state probability of failure in ordinary conditions (i.e., at the steady-state portion of the "bathtub curve"). If the possible failure is attributed, for example, to random vibrations, then the following Steinberg's formula can be used to assess the MTTF:

$$\tau = C\sigma_r^{-m/2}. \tag{5.14}$$

Here σ_r is the mechanical stress at the resonance frequency, and C and m are material (structural) parameters that can be established by accelerated life

Probabilistic Assessment of an Aerospace Mission Outcome 117

testing. The above formula reflects an assumption that the MTTF is determined, for the given material and structure, by the square root of the resonant stress. The predicted failure rate is, therefore,

$$\lambda = \frac{1}{C}\sigma_r^{m/2}.$$

(5.15)

5.6 Human Performance Failure Rate

By analogy with how the failure rate for a piece of equipment is determined, one could use the condition (5.12) to establish the similar relationship for the human performance. We view the process of testing and training of a human on a simulator as a sort of an accelerated test setup for a vehicle operator. From (5.1) we have, for $F = F_0$ (i.e., using patent law terminology), for a human of the ordinary skills in the vehicular "art," the following formula for the probability of nonfailure, when a navigator is being tested or trained on a flight simulator:

$$P^h(G) = P_0 \exp\left(1 - \frac{G^2}{G_0^2}\right).$$

(5.16)

Then the probability of his/her failure is

$$Q_h(G) = 1 - P^h(G) = 1 - P_0 \exp\left(1 - \frac{G^2}{G_0^2}\right)$$

(5.17)

and

$$\tau = \frac{1}{\lambda} = \frac{\tau_0}{Q_h(G)} = \frac{\tau_0}{1 - P_0 \exp\left(1 - \frac{G^2}{G_0^2}\right)}.$$

(5.18)

This formula can be employed to run an ALT procedure on a simulator, using the elevated MWL level G as the stimulus factor, to the same extent as the elevated absolute temperature is used, when Arrhenius equation is employed, to accelerate failures in the relationship (5.11). The parameters G_0, τ_0, and P_0 should be viewed as empirical parameters that could be determined from the relationship (5.18) as a result of testing at different MWL levels G for many individuals and by evaluating the corresponding MTTF τ.

Note, that as far as steady-state condition is concerned, we use the simplest, exponential, distribution for the evaluation of the probability P_0, while in our general mission success and safety concept, reflected by Equation (5.10), we use a more general and more flexible Weibull distribution. Since there are three experimental parameters in the relationship (5.18) that have to be determined, one needs three independent equations to determine these parameters. If the tests on a simulator are being conducted for three groups of individuals at three MWL levels G_1, G_2, and G_3, and their performance is measured by recording three times-to-failure, τ_1, τ_2, and τ_3, then the G_0 value can be obtained from the following transcendental equation:

$$\left(1 - \frac{\tau_1}{\tau_2}\right)\left[\exp\left(1 - \frac{G_3^2}{G_0^2}\right) - \frac{\tau_2}{\tau_3}\exp\left(1 - \frac{G_2^2}{G_0^2}\right)\right]$$

$$-\left(1 - \frac{\tau_2}{\tau_3}\right)\left[\exp\left(1 - \frac{G_2^2}{G_0^2}\right) - \frac{\tau_1}{\tau_2}\exp\left(1 - \frac{G_1^2}{G_0^2}\right)\right] = 0. \tag{5.19}$$

One could easily check that this equation is always fulfilled for $G_1 = G_2 = G_3 = G_0$. It is noteworthy that, as has been determined previously on the basis of Table 5.1 data, testing is not (and should not be) conducted at MWL levels essentially higher than threefold higher than the normal MWL, otherwise a "shift" in the mode of failure (i.e., misleading results) is likely. In other words, the accelerated test conditions should be indeed accelerated ones, and have to be reasonably high, but should not be unrealistically/unreasonably high. We are all still human, not superhuman, and, even an experienced, young-yet-mature, competent, and well-trained individual (like Captain Sullenberger, for instance) cannot cope with an exceptionally and perhaps unrealistically high workload. After the normal (most likely) MWL G_0 is evaluated, the probability of nonfailure at normal MWL conditions can be found as

$$P_0 = \frac{1 - \dfrac{\tau_1}{\tau_2}}{\exp\left(1 - \dfrac{G_2^2}{G_0^2}\right) - \dfrac{\tau_1}{\tau_2}\exp\left(1 - \dfrac{G_1^2}{G_0^2}\right)} = \frac{1 - \dfrac{\tau_2}{\tau_3}}{\exp\left(1 - \dfrac{G_3^2}{G_0^2}\right) - \dfrac{\tau_2}{\tau_3}\exp\left(1 - \dfrac{G_2^2}{G_0^2}\right)}, \tag{5.20}$$

and the time τ_0, can be then determined, if necessary, as

$$\tau_0 = \tau_1\left[1 - P_0\exp\left(1 - \frac{G_1^2}{G_0^2}\right)\right] = \tau_2\left[1 - P_0\exp\left(1 - \frac{G_2^2}{G_0^2}\right)\right] = \tau_3\left[1 - P_0\exp\left(1 - \frac{G_3^2}{G_0^2}\right)\right]. \tag{5.21}$$

As evident from the formulas (5.19–5.21), the G_0 value can be found in a single way from the formula (5.19), the P_0 value can be found in two ways using the formulas (5.20), and the τ_0 value can be found in three ways using the formulas (5.21). This circumstance should be used to check the accuracy in

Probabilistic Assessment of an Aerospace Mission Outcome

determining these values. For the analysis based on Equation (5.10), only the P_0 value is needed. We would like to point out also that, although a minimum of three levels of the MWL are needed to determine the parameters G_0, τ_0, and P_0, it is advisable that tests at many more MWL levels (still within the range $\dfrac{G}{G_0} = 1-3$) are conducted, so that the accuracy in the prediction could be assessed. After the parameters G_0, τ_0, and P_0 are found, the failure rate can be determined as a function of the MWL level from the formula (5.18):

$$\lambda = \frac{1}{\tau_0}\left[1 - P_0 \exp\left(1 - \frac{G^2}{G_0^2}\right)\right].$$

(5.22)

The nominal (normal, ordinary, specified) failure rate is, therefore,

$$\lambda = \frac{1 - P_0}{\tau_0}.$$

(5.23)

5.7 Weibull Law

We use the Weibull law to evaluate the time effect (aging, degradation) on the performance of both the equipment (instrumentation), considering the combined effect of the hardware and software, and the "human-in-the-loop." It is a two-parametric distribution with the probability distribution function

$$F(t) = e^{-(\lambda t)^\beta}$$

(5.24)

where the failure rate λ is related to the scale parameter η of the distribution as $\eta = \dfrac{1}{\lambda}$, and the MTTF $\bar{t}$ and the standard deviation σ_t of the time-to-failure t can be found as

$$\bar{t} = \eta\Gamma\left(1 + \frac{1}{\beta}\right), \quad \sigma_t = \eta\sqrt{\Gamma\left(1 + \frac{2}{\beta}\right) - \Gamma^2\left(1 + \frac{1}{\beta}\right)}.$$

(5.25)

Here,

$$\Gamma(\alpha) = \int_0^\infty x^{\alpha-1} e^{-x} dx$$

(5.26)

is the gamma-function. The probability density distribution function can be obtained, if needed, from (5.24) by differentiation:

$$f(t) = \lambda\beta(\lambda t)^{\beta-1} e^{-(\lambda t)^\beta}.$$

(5.27)

Example 5.1. Let, for instance, the duration of a particular vehicular mission be 24h, and the vehicle spends equal times at each of the six segments (so that $t_i = 4$h at the end of each segment), the failure rates of the equipment and the human performance are independent of the environmental conditions and are $\lambda = 8 \times 10^{-4}\,1/$h, the shape parameter in the Weibull distribution in both cases is $\beta = 2$ (Rayleigh distribution), the HCF ratio $\dfrac{F^2}{F_0^2}$ is $\dfrac{F^2}{F_0^2} = 8$ (so that $\dfrac{F}{F_0} = 2.828$), the probability of human nonfailure at ordinary conditions is $P_0 = 0.9900$, and the MWL G_i^2/G_0^2 ratios are given versus the probability q_i of occurrence of the environmental conditions in Table 5.2. Table 5.2 data presume that about 95% of the mission time occurs in ordinary conditions. The computations of the probabilities of interest are also carried out in Table 5.2. We obtain the following:

$$P_i^e = \exp\left[-\left(\lambda t_i\right)^2\right] = \exp\left[-\left(8 \times 10^{-4} \times 4\right)^2\right] = 0.99999,$$

$$P_i^h = P_0 \bar{P}_i \exp\left[-\left(\lambda t_i\right)^2\right] = 0.9900 \times 0.99999 \bar{P}_i = 0.99 \bar{P}_i$$

and

$$\sum_{i=1}^{n} q_i P_i^e\left(t_i\right) P_i^h\left(t_i\right) = 0.9900,$$

which is the probability of the mission nonfailure. The overall probability of mission failure is, therefore,

$$Q = 1 - \sum_{i=1}^{n} q_i P_i^e\left(t_i\right) P_i^h\left(t_i\right) = 1 - 0.9900 = 0.01 = 1\%.$$

TABLE 5.2

Calculated Probabilities of Mission Failure

i	1	2	3	4	5	6
$q_i,\%$	95.30	3.99	0.50	0.10	0.06	0.05
G_i/G_0	1	1.4142	1.7324	2.0000	2.2361	2.4495
$\bar{P}_i$	1	0.9991	0.9982	0.9978	0.9964	0.9955
P_i^h	0.9900	0.9891	0.9882	0.9878	0.9864	0.9855
$P_i^e P_i^h$	0.9900	0.9891	0.9882	0.9878	0.9864	0.9855
$q_i P_i^e P_i^h$	0.9435	0.0395	0.0049	0.0010	0.0006	0.0005

5.8 Imperfect Human versus Imperfect Instrumentation: Some Short-Term Predictions

The concept based on the formula (5.10) and addressed in Sections 5.2–5.6 is suitable for the design of the hardware and the software, for making long-term assessments and strategic decisions, and for planning a certain vehicular mission before this mission actually commences. There are, however, extraordinary situations, when the navigator has to make a decision on a short-term, even on an emergency, basis during the actual fulfillment of the mission. Many examples (problems) associated with the application of probabilistic methods to quantify the combined effect of the human-equipment-environment interaction were provided in Chapter 2. Here are several additional and/or somewhat differently formulated examples.

> **Example 5.2.** The probability that the particular environmental conditions will be detrimental for the vehicle safety (say, the probability of exceeding a certain probability level) is p. The probability that these conditions are detected by the available navigation equipment, adequately processed and delivered to the navigator in due time is p_1. But the navigator is not perfect either, and the probability that he/she misinterprets the obtained information from the navigation instrumentation is p_2. If this happens, the navigator can either launch a false alarm (take inappropriate and unnecessary corrective actions), or conclude that the weather conditions are acceptable and make an inappropriate go-ahead decision. The navigator receives n messages from the navigation equipment during his/her watch. What is the probability that at least one of the messages will be assessed incorrectly?
>
> The hypotheses about a certain message are as follows: H_1 = the weather conditions are unacceptable, so that the corrective actions are necessary; H_2 = the weather conditions are acceptable and therefore no corrective actions are needed. The probability that a message is misinterpreted is
>
> $$P = p(1-p_1)+(1-p)p_2. \tag{5.28}$$
>
> Then the probability that at least one message out of n is misinterpreted is
>
> $$Q = 1-(1-P)^n. \tag{5.29}$$
>
> Clearly, $Q \to 1$, when $n \to \infty$. The previous formulas indicate that the outcome depends on both the equipment (instrumentation) performance and the human ability to correctly interpret the obtained information. The formula (5.29) can be used particularly to assess the effect of the human fatigue on his/her ability to interpret correctly the obtained messages. Let, for instance, $n = 100$ (the navigator receives 100 messages during his/her watch), and $p = 1$: the forecast environmental conditions that the vehicle will encounter will certainly cause an accident and should

be avoided. So, the instrumentation did not fail, and the probability p_1 that the navigator obtained this information and that the information has been delivered in a timely fashion is $p_1 = 0.999$. Let the probability that the navigator interprets the information incorrectly be, say, only $p_2 = 0.01 = 1\%$. Then $P = 0.001$ and $Q = 0.0952$. Thus, the probability that one message could be misinterpreted is as high as 9.5%. If the equipment is not performing adequately and the probability p_1 is only, say, $p_1 = 0.95$, then $P = 0.05$ and $Q = 0.9941$: one of the messages from the navigation equipment will be most certainly misinterpreted. Thus, we conclude that the performance and the accuracy of the instrumentation are as important as the human factor.

Example 5.3. The probability that the instrumentation does not fail during the time T of the fulfillment of a certain segment of a mission is p_1. The probability that the human "does not fail" (i.e., receives and interprets the obtained information correctly [does not make any error]) during this time is p_2. It has been established that a certain (nonfatal though) accident has occurred during the time of the fulfillment of this segment of the mission. What is the probability that the accident occurred because of the equipment failure? Four hypotheses were possible before the accident actually occurred: H_0 = the equipment did not fail and the human did not make any error; H_1 = the equipment failed, but no human error occurred; H_2 = the equipment did not fail, but the human made an error; H_3 = the equipment failed and the human made an error. The probabilities of these hypotheses are

$$P(H_0) = p_1 p_2 \; ; P(H_1) = (1 - p_1) p_2 \; ; P(H_2) = p_1 (1 - p_2) ; P(H_3) = (1 - p_1)(1 - p_2).$$

The conditional probabilities of the event A "the accident occurred" are

$$P(A/H_0) = 0, \quad P(A/H_1) = P(A/H_2) = P(A/H_3) = 1.$$

By applying Bayes' formula,

$$P(H_i/A) = \frac{P(H_i)P(A/H_i)}{\displaystyle\sum_{i=1}^{n} P(H_i)P(A/H_i)}, \quad i = 1, 2, \ldots, n,$$

we obtain the following expression for the probability that only the equipment failed:

$$P(H_i/A) = \frac{(1 - p_1) p_2}{(1 - p_1) p_2 + p_1 (1 - p_2) + (1 - p_1)(1 - p_2)} = \frac{(1 - p_1) p_2}{1 - p_1 p_2}. \quad (5.30)$$

Clearly, if the equipment never fails ($p_1 = 1$), then $P = 0$. But, if the equipment is very unreliable ($p_1 = 0$), then $P = p_2$: the probability that the equipment fails is equal to the probability that the operator did not make an

Probabilistic Assessment of an Aerospace Mission Outcome

error. If the probabilities p_1 and p_2 are equal ($p_1 = p_2 = p$), then $P = \dfrac{p}{1+p}$ is the probability that either the equipment failed or the human made an error. For very reliable equipment and a next-to-perfect operator (human) ($p = 1$), $P = 0.5$: the probability that only the equipment failed is 0.5. For very unreliable equipment and very "imperfect" human ($p = 0$), we obtain $P = 0$: it is quite likely that both the equipment failed and the human made an error.

Example 5.4. The assessed probability that a certain segment of a mission will be accomplished successfully, provided that the environmental conditions are favorable, is p_1. This probability will not change even in unfavorable environmental conditions, if the navigation equipment is adequate and functions properly. If, however, the equipment (instrumentation) is not perfect, then the probability of safe fulfillment of the given segment of the mission is only $p_2 \vartriangleleft p_1$. It has been established that the probability of failure-free functioning of the navigation equipment is p_*. It is known also that in this region of the navigation space unfavorable navigation conditions are observed at the given time of the year in $k\%$ of the time. What is the probability of the successful accomplishment of the mission in any environmental conditions? What is the probability that the navigator used the equipment, if it is known that the mission has been accomplished successfully?

The probability of the hypothesis H_1 "the environmental conditions are favorable" is $P(H_1) = 1 - \dfrac{k}{100}$. The probability of the hypothesis H_2 "the environmental conditions are unfavorable" is $P(H_2) = \dfrac{k}{100}$. The conditional probability $P(A/H_1)$ of the event A "the navigation is safe" when the environmental conditions are favorable is $P(A/H_1) = p_1$. The conditional probability $P(A/H_2)$ of the event A "the navigation is safe" when the environmental conditions are unfavorable can be determined as

$$P(A/H_2) = p_* p_1 + (1 - p_*)p_2,$$

so that the sought probability of accident-free navigation on the given segment is

$$P(A) = \left(1 - \frac{k}{100}\right)p_1 + \frac{k}{100}\left[p_* p_1 + (1 - p_*)p_2\right] = p_1 - \frac{k}{100}(p_1 - p_2)(1 - p_*).$$

If it is known that the mission has been accomplished successfully despite unfavorable environmental conditions, then

$$P(A/H_2) = \frac{\dfrac{k}{100}\left[p_* p_1 + (1 - p_*)p_2\right]}{P(A)} = \frac{\dfrac{k}{100}\left[p_* p_1 + (1 - p_*)p_2\right]}{p_1 - \dfrac{k}{100}(p_1 - p_2)(1 - p_*)}. \tag{5.31}$$

124 Human-in-the-Loop

Let, for instance, $p_1 = 1.0$, $p_2 = 0.95$, $p_* = 0.98$, $k = 80$. Then $P(A) = 0.9992$, $P(A/H_2) = 0.7998$.

So, the probability of the successful accomplishment of the mission is 0.9992, and the probability that the navigator used the navigation instrumentation/equipment that enabled him/her to accomplish the mission successfully is 0.7998, otherwise the mission would have failed.

Example 5.5. The q_i values for the wave conditions in the North Atlantic in the region between 50° and 60° North Latitude are shown in Table 5.3 versus wave heights of 3% significance (wave heights of 3% significance means that 97% of the waves are characterized by the heights below the $h_{3\%,m}$ level, and 3% have the height exceeding this level): Two sources of information predict a particular q_i value at the next segment of the route with different probabilities p_1 and p_2. What is the likelihood that the first source is more trustworthy than the second one?

Let A be the event "the first forecaster is right," $\bar{A}$ be the event "the first forecaster is wrong," B be the event "the second forecaster is right," and $\bar{B}$ be the event "the second forecaster is wrong." So, we have $P(A) = p_1$ and $P(B) = p_2$. Since the two forecasters (sources) made different predictions, the event $A\bar{B} + \bar{A}B$ took place. The probability of this event is

$$P(A\bar{B} + \bar{A}B) = P(A\bar{B}) + P(\bar{A}B) = P(A)P(\bar{B}) + P(\bar{A})P(B) = p_1(1-p_2) + (1-p_1)p_2.$$

The first forecaster will be more trustworthy if the event $A\bar{B}$ takes place. The probability of this event is

$$P(A\bar{B}) = \frac{p_1(1-p_2)}{p_1(1-p_2) + (1-p_1)p_2} = \frac{1}{1 + \dfrac{1-p_1}{1-p_2}\dfrac{p_2}{p_1}}. \tag{5.32}$$

The relationship (5.32) is computed in Table 5.4. Clearly, $P(A\bar{B}) = 0.5$, if $p_1 = p_2 = p$; $P(A\bar{B}) = 1.0$, if $p_1 = 1$ and $p_2 \neq 1$; $P(A\bar{B}) = 0$, if $p_1 \neq 1$ and $p_2 = 1$. Other Table 4.4 data are not counterintuitive, but this table quantifies the role of the two mutually exclusive forecasts.

TABLE 5.3

Probability of Encounter of the Environmental Conditions of the Given Level of Severity

$h_{3\%,m}$	3	6	9	12	15	18
q_i	0.1500	0.0501	0.0092	0.000876	0.0000437	0.00000115

TABLE 5.4

Calculated Trustworthiness of Weather Forecast

p_2 \ p_1	0.1	0.2	0.4	0.6	0.8	0.9
0.1	0.500	0.692	0.857	0.931	0.973	0.988
0.2	0.308	0.500	0.727	0.857	0.941	0.973
0.4	0.143	0.273	0.500	0.692	0.857	0.931
0.6	0.069	0.143	0.308	0.500	0.727	0.857
0.8	0.027	0.059	0.143	0.273	0.500	0.692
0.9	0.012	0.027	0.069	0.143	0.308	0.500

5.9 Most Likely MWL

Cognitive overload has been recognized as a significant cause of error in aviation, and therefore, measuring the MWL has become a key method of improving safety. There is an extensive published work in the psychological literature devoted to the measurement of MWL, both in military and in civil aviation (see, for instance, [4–25]). A pilot's MWL can be measured using subjective ratings or objective measures. The subjective ratings during simulation tests can be in the form of periodic inputs to some kind of data collection device that prompts the pilot to enter a number between 1 and 7 (for example) to estimate the MWL every few minutes. Another possible approach is postflight paper questionnaires.

There are some objective measures of MWL, such as heart rate variability. It is easier to measure the MWL in a flight simulator than in actual flight conditions. In a real airplane, one would probably be restricted to using postflight subjective (questionnaire) measures, since one would not want to interfere with the pilot's work. An aircraft pilot faces numerous challenges imposed by the need to control a multivariate lagged system in a heterogeneous multitask environment. The time lags between critical variables require predictions and actions in an uncertain world. The interrelated concepts of situation awareness and MWL are central to aviation psychology. The major components of situation awareness are spatial awareness, system awareness, and task awareness. Each of these three components has real-world implications: spatial awareness—for instrument displays, system awareness—for keeping the operator informed about actions that have been taken by automated systems, and task awareness—for attention and task management.

Task management is directly related to the level of the MWL, as the competing "demands" of the tasks for attention might exceed the operator's resources—his/her "capacity" to adequately cope with the "demands" imposed by the MWL. In modern military aircraft, complexity of information, combined with time stress, creates difficulties for the pilot under combat conditions, and the first step to mitigate this problem is to measure and manage MWL [5]. Although there is no universally accepted definition of the MWL and how it should/could be evaluated, there is a consensus that suggests that MWL can be conceptualized as the interaction between the structure of systems and tasks, on the one hand, and the capabilities, motivation, and state of the human operator, on the other. More specifically, MWL could be defined as the "cost" that an operator incurs as tasks are performed. Given the multidimensional nature of MWL, no single measurement technique can be expected to account for all the important aspects of it. Current research efforts in measuring MWL use psycho-physiological techniques, such as electroencephalographic, cardiac, ocular, and respiration measures in an attempt to identify and predict MWL levels. Measurement of cardiac activity has been a useful physiological technique employed in the assessment of MWL, both from tonic variations in heart rate and after treatment of the cardiac signal.

5.10 Most Likely HCF

The HCF includes the person's professional experience; qualifications; capabilities; skills; training; sustainability; ability to concentrate; ability to operate effectively, in a "tireless" fashion, under pressure, and, if needed, for a long period of time; ability to act as a "team player"; swiftness of reaction—that is, all of the qualities that would enable him/her to cope with high MWL. In order to come up with a suitable figures of merit (FOM) for the HCF, one could rank each of the above and other qualities on a scale from 1 to 10, and calculate the average figures of merit (FOM) for each individual.

5.11 Conclusions

The following conclusions can be drawn from the carried out analyses:

- A DEPDF of the EVD type is introduced to characterize and to quantify the likelihood of a human failure to perform his/her duties when operating a vehicle (a car, an aircraft, a boat, etc.). This function is applied to assess a mission success situation.
- We have shown how some methods of the classical probability theory could be employed to quantify the role of the human factor in the situation in question. We show that if highly reliable equipment

is used, the mission could still be successful, even if the HCF is not very high.

- The suggested PRA approach complements the existing system-related and human psychology–related efforts, and, most importantly, bridges the gap between the three critical areas responsible for the system performance: reliability engineering, vehicular technologies, and human factor.

- Plenty of additional PRM analyses and human psychology–related effort will be needed, of course, to make the guidelines based on the suggested concept practical for particular applications.

- These applications might not even be necessarily in the vehicular technology domain, but in many other areas and systems (forensic, medical, etc.), where a human interacts with equipment and instrumentation, and operates in conditions of uncertainty.

- Although the approach is promising and fruitful, further research, refinement, and validation would be needed, of course, before the model could become practical. The suggested model, after appropriate sensitivity analysis is carried out, might be used when developing guidelines for personnel training and/or when there is a need to decide if the existing navigation instrumentation is adequate in extraordinary safety-in-air situations, or if additional and/or more advanced equipment should be developed and installed.

- The initial numerical data based on the suggested model make physical sense and are in satisfactory (qualitative) agreement with the existing practice. It is important to relate the model expressed by the basic Equation (5.1) to the existing practice, on one hand, and to review the existing practice from the standpoint of this model on the other.

References

1. E. Suhir, *Applied Probability for Engineers and Scientists*, McGraw-Hill, New York, 1997.
2. E. Suhir, *Adequate Underkeel Clearance (UKC) for a Ship Passing a Shallow Waterway: Application of the Extreme Value Distribution (EVD)*, Rio-de-Janeiro, Brazil, OMAE2001/S&R-2113, 2001.
3. E. Suhir, *Helicopter Landing Ship (HLS): Undercarriage Strength and the Role of the Human Factor*, Honolulu, Hawaii, OMAE, 2009. See also ASME Transactions, *OMAE Journal*, February 2010.
4. E. Suhir, "Probabilistic Modeling of the Role of the Human Factor in the Helicopter-Landing-Ship (HLS) Situation," *International Journal of Human Factor Modeling and Simulation (IJHFMS)*, vol. 1, No. 3, 2010, pp. 1–8.

5. E. Suhir and R. Mogford,"Two-Men-in-a-Cockpit: Casualty Likelihood if One Pilot Becomes Incapacitated," *Journal of Aircraft*, vol. 48, No. 4, 2011, pp. 1309–1314. See also AIAA ATIO/ISSMO Conference: Control ID:822771, Renaissance Worthington Hotel, Fort Worth, Texas, September 13–15, 2010.
6. D.C. Foyle and B.L. Hooey, *Human Performance Modeling in Aviation*, CRC Press, Boca Raton, FL, 2008.
7. E. Svensson, M. Angelborg-Thanderz, and L. Sjoeberg, "Mission Challenge, Mental Workload and Performance in Military Aviation," *Aviation, Space, and Environmental Medicine*, vol. 64, No. 11, 1993, pp. 985–991.
8. T.C. Hankins and G.F. Wilson, "A Comparison of Heart Rate, Eye Activity, EEG, and Subjective Measures of Pilot Mental Workload During Flight," *Aviation, Space, and Environmental Medicine*, vol. 69, 1998, pp. 360–367.
9. K.A. Greene, K.W. Bauer, M. Kabrisky, S.K. Rogers, and G.F. Wilson, "Estimating Pilot Workload Using Elman Recurrent Neural Networks: A Preliminary Investigation." In C.H. Dagli, et al. (eds.), *Intelligent Engineering Systems through Artificial Neural Networks*, vol. 7, ASME Press, New York, 1997.
10. J.A. East, K.W. Bauer, and J.W. Lanning, "Feature Selection for Predicting Pilot Mental Workload: A Feasibility Study," *International Journal of Smart Engineering System Design*, vol. 4, 2002, pp. 183–193.
11. J.B. Noel, "Pilot Mental Workload Calibration." MS thesis, School of Engineering, Air Force Institute of Technology, Wright-Patterson AFB, OH, March 2001.
12. A.W. Gaillard, "Comparing the Concepts of Mental Load and Stress," *Ergonomics*, vol. 36, 1993, pp. 991–1005.
13. A.F. Kramer, "Physiological Metrics of Mental Workload: A Review of Recent Progress." In P. Ullsperger (ed.), *Mental Workload*, Bundesanstaltfür Arbeitmedizin, Berlin, 1993.
14. L.R. Fournier, G.F. Wilson, and C.R. Swain, "Electrophysiological, Behavioral and Subjective Indexes of Workload When Performing Multiple Tasks: Manipulation of Task Difficulty and Training," *International Journal of Psychophysiology*, vol. 31, 1999, pp. 277–283.
15. P. Ullsperger, A. Metz, and H. Gille, "The P300 Component of the Event-Related Brain Potential and Mental Effort,"*Ergonomics*, vol. 31, 1988, pp. 1127–1137.
16. G.F. Wilson, P. Fullerkamp, and I. Davis, "Evoked Potential, Cardiac, Blink and Respiration Measures of Pilot's Workload in Air-to-Ground Missions," *Aviation, Space and Environmental Medicine*, vol. 65, 1994.
17. J.B. Brookings, G.F. Wilson, and C.R. Swain, "Psycho-Physiological Responses to Changes in Workload During Simulated Air Traffic Control," *Biological Psychology*, vol. 42, 1996, pp. 361–377.
18. T.C. Hankins and G.F. Wilson, "A Comparison of Heart Rate, Eye Activity EEG and Subjective Measures of Pilot Mental Workload During Flight," *Aviation, Space and Environmental Medicine*, vol. 69, 1998, pp. 2713–2730.
19. A.H. Roscoe, "Heart Rate as a Psycho-Physiological Measure for In-Flight Workload Assessment," *Ergonomics*, vol. 36, 1993, pp. 1055–1062.
20. R.W. Backs, "Going Beyond Heart Rate: Autonomic Space and Cardiovascular Assessment of Mental Workload," *International Journal of Aviation Psychology*, vol. 5, 1995, pp. 25–48.
21. A.J. Tattersall and G.R. Hockey, "Level of Operator Control and Changes in Heart Rate Variability During Simulated Flight Maintenance," *Human Factors*, vol. 37, 1995.

22. P.G. Jorna, "Heart Rate and Workload Variations in Actual and Simulated Flight," *Ergonomics*, vol. 36, 1993, pp. 1043–1054.
23. J.A. Veltman and A.W. Gaillard, "Physiological Indices of Workload in a Simulated Flight Task," *Biological Psychology*, vol. 42, 1996, pp. 323–342.
24. M.R. Endsley and M.D. Rogers, "Distribution of Attention, Situation Awareness and Workload in a Passive Air Traffic Control Task: Implications for Operational Errors and Automation," *Air Traffic Control Quarterly*, vol. 6, No. 1, 1988.
25. C.D. Wickers, A.S. Mavor, and J.P. McGee (eds.), "Flight to the Future: Human Factors in Air Traffic Control," Panel on human factors in air traffic control automation, Committee on human factors, http://www.nap.edu/catalog/5493html, 1997.
26. M.A. Staal, "Stress, Cognition, and Human Performance: A Literature Review and Conceptual Framework," NASA/TM-2004-212824, November 2003.
27. L.L. Murphy, K. Smith, and P.A. Hancock, "Task Demand and Response Error in a Simulated Air Traffic Control Task: Implications for ABInitio Training," *International Journal of Applied Aviation Studies*, vol. 4, No. 1, 2004.
28. E. Suhir, "How to Make a Device into a Product: Accelerated Life Testing, Its Role, Attributes, Challenges, Pitfalls and Interaction with Qualification Testing." In E. Suhir, C.P. Wong, and Y.C. Lee (eds.), *Micro- and Opto-Electronic Materials and Structures: Physics, Mechanics, Design, Packaging, Reliability*, Springer, 2007.
29. E. Suhir, "Probabilistic Design for Reliability," *ChipScale Reviews*, vol. 14, No. 6, 2010.
30. E. Suhir and R. Mahajan, "Are Current Qualification Practices Adequate?" *Circuit Assembly*, April 2011.

6

The "Miracle-on-the-Hudson" Event: Quantitative Aftermath

6.1 Summary

Application of the quantitative probabilistic risk analysis (PRA) concept should complement in various human-in-the-loop (HITL) situations, whenever feasible and possible, the existing vehicular psychology practices, which are typically qualitative *a posteriori* statistical assessments. A PRA approach based on the double-exponential probability distribution function (DEPDF) is suggested as a suitable quantitative technique for assessing the probability of the human nonfailure in an off-normal flight situation. The (typically long-term) human capacity factor (HCF) is introduced in this distribution and considered along with the elevated short-term mental workload (MWL) that the human (pilot) has to cope with in a highly off-normal (emergency) situation.

The famous 2009 US Airways "Miracle-on-the-Hudson" successful landing (ditching) and the infamous 1998 Swissair "UN-shuttle" disaster are chosen to illustrate the usefulness and fruitfulness of the approach in the problem in question. It is argued particularly that it was the exceptionally high HCF of the US Airways crew and especially that of Captain Sullenberger (the author was told that some pilots disagree with such an assessment) that made what seemed to be, at the first glance, a "miracle," the reality. It is shown also, as a comparison, that the highly professional and, in general, highly qualified Swissair crew exhibited inadequate performance (quantified in our hypothetical analysis as a relatively low HCF level) in the off-normal situation they encountered. The crew made several fatal errors and, as a result, crashed the aircraft. In addition to the DEPDF-based approach, it is shown that the probability of safe landing can be modeled and predicted by comparing the (random) operation time (which, like in the previously addressed helicopter-landing-ship problem, consists of the decision-making time and the actual landing time) with the "available" (objective) time needed for landing.

It is concluded that the developed formalisms, after trustworthy input data are obtained (using, e.g., flight simulators, or applying Delphi method, or otherwise) might be applicable even beyond the considered situation and even beyond the vehicular domain. The developed methodology can be employed in various HITL situations, when an highly effective short-term human performance is imperative, and the ability to quantify it is, therefore, highly desirable. It is concluded that, although the generated hypothetical numbers that characterize human performance make physical sense, it is the approach, and not these numbers, that is the main merit of the analysis in this chapter.

6.2 Introduction

In this chapter, the DEPDF-based model is applied for the evaluation of the likelihood of a human nonfailure in an emergency situation. The famous 2009 Miracle-on-the-Hudson event and the infamous 1998 UN-shuttle disaster are used to illustrate the substance and fruitfulness of the employed MWL/HCF approach. We try to shed probabilistic light on these two events. As far as the "miracle" is concerned, we intend to provide tentative quantitative assessments of why such a "miracle" could have actually occurred, and what had been and had not been indeed a "miracle" in this incident: a divine intervention, a perceptible interruption of the laws of nature, or merely a wonderful and rare occurrence that was due to a heroic act of the aircraft crew and especially of its Captain Sullenberger, the lead "miracle worker" in the incident. As to the UN-shuttle crash, we are going to demonstrate that the crash might have occurred because of the low HCF of the aircraft crew in an off-normal situation that they had encountered and that was, in effect, much less demanding/challenging than the Miracle-on-the-Hudson situation. It should be emphasized that all of the HCF ratings are absolutely hypothetical: it could be this way, if anyone would have tried to apply quantitative assessments of the HCF for the main characters of the two events.

In order to justify what I did in this chapter, I brought in the second epigraph to this book: "There are things in this world, far more important than the most splendid discoveries—it is the methods by which they were made," from Gottfried Leibnitz, the great German mathematician, the one who, independently of Newton, "discovered" calculus, the mathematical study of continuous change. By quoting Leibnitz, I tried to deliver the major message of this book: quantifications of the pilot's HCF could be done differently, but they should have been done.

6.3 Miracle-on-the-Hudson Event, and the Roles of MWL and HCF

The HCF versus MWL concept considers elevated (off-normal) random relative HCF and MWL levels with respect to the ordinary (normal, pre-established) deterministic HCF and MWL values. These values are or should be established in one way or another on the basis of the existing human psychology practices in avionics. As is known, various interrelated concepts of situation awareness and MWL ("demand") are central to today's aviation psychology. Cognitive (mental) overload has been recognized as a significant cause of error in aviation. The MWL is directly affected by the challenges that a navigator faces, when controlling the vehicle in a complex, heterogeneous, multitask, and often uncertain and harsh environment. Such an environment includes numerous different and interrelated concepts of situation awareness: spatial awareness for instrument displays; system awareness for keeping the pilot informed about actions that have been taken by automated systems; and task awareness that has to do with the attention and task management. The time lags between critical variables require predictions and actions in an uncertain world.

The MWL depends on the operational conditions and on the complexity of the mission. MWL has to do, therefore, with the significance of the long- or short-term task. Task management is directly related to the level of the MWL, as the competing "demands" of the tasks for attention might exceed the operator's resources—his/her "capacity" to adequately cope with the "demands" imposed by the MWL. Measuring the MWL has become a key method of improving aviation safety. There is an extensive published work in the psychological literature devoted to the measurement of the MWL in aviation, both military and commercial. A pilot's MWL can be measured using subjective ratings and/or objective measures. The subjective ratings during failure-oriented-accelerated testing (FOAT) (simulation tests) can be, for example, after the expected failure is defined, in the form of periodic inputs to some kind of data collection device that prompts the pilot to enter a number between 1 and 10 (for example) to estimate the MWL every few minutes. There are some objective MWL measures, such as, for example, heart rate variability. Another possible approach uses postflight paper questionnaires. It is easier to measure the MWL on a flight simulator than in actual flight conditions. In a real aircraft, one would probably be restricted to using postflight subjective (questionnaire) measurements, since one would not want to interfere with the pilot's work.

Given the multidimensional nature of MWL, no single measurement technique can be expected to account for all of the important aspects of it. In modern military aircraft, complexity of information, combined with time stress, create difficulties for the pilot under combat conditions, and the first step to mitigate this problem is to measure and manage the MWL. Current research efforts in measuring MWL use psycho-physiological techniques, such as, for example, electroencephalographic, cardiac, ocular, and respiration measures in an attempt to identify and predict MWL levels. Measurement of cardiac activity has been a useful physiological technique employed in the assessment of MWL, both from tonic variations in heart rate and after treatment of the cardiac signal.

As to the HCF, as has been indicated in the previous chapters, it includes, but might not be limited to, the following major qualities that would enable a professional human to successfully cope with an elevated off-normal MWL: psychological suitability for a particular task; professional experience and qualifications; education, both special and general; relevant capabilities and skills; level, quality, and timeliness of training; performance sustainability (consistency, predictability); independent thinking and independent acting, when necessary; ability to concentrate; ability to anticipate; self-control and the ability to act in cold blood in hazardous and even life-threatening situations; mature (realistic) thinking; ability to operate effectively under pressure, and particularly under time pressure; ability to operate effectively, when necessary, in a tireless fashion, for a long period of time (tolerance to stress); ability to act effectively under time pressure and make well-substantiated decisions in a short period of time; team-player attitude, when necessary; and swiftness in reaction, when necessary. These and other qualities are certainly of different importance in different situations. It is clear also that different individuals possess these qualities in different degrees (i.e., differently psychologically suitable for a particular task).

Both MWL and HCF could be time dependent. In order to come up with a suitable figures-of-merit (FOM) for the HCF, one could rank, similarly to the MWL estimates for particular situations or missions, the above and perhaps other qualities on a scale from, say, 1 to 4, and calculate the average FOM for each individual and particular task (see, e.g., Tables 6.5, 6.6, and 6.8). It goes without saying that the MWL and HCF should be measured using the same units.

6.4 Double-Exponential Probability Distribution Function

Different PRA approaches can be used in the analysis and optimization of the interaction of the MWL and HCF. When the MWL and HCF characteristics are treated as deterministic characteristics, a safety factor $SF = \dfrac{HCF}{MWL}$ can be used. When both MWL and HCF are random variables, the safety

The "Miracle-on-the-Hudson" Event

factor can be determined as the ratio $SF = \dfrac{\prec SM \succ}{S_{SM}}$ of the mean value $\prec SM \succ$ of the random safety margin SM = HCF − MWL to its standard deviation S_{SM}. When the capacity–demand ("strength–stress") interference model is used, the HCF can be viewed as the capacity (strength) and the MWL as the demand (stress), and their overlap area could be considered as the potential (probability) of possible human failure.

The capacity and the demand distributions can be steady state or transient (i.e., their mean values can move toward each other when time progresses), and/or the MWL and HCF curves can get spread over larger areas.

Yet another PRA approach is to use a single distribution that accounts for the roles of the HCF and MWL, when these (random) characteristics deviate from (are higher than) their (deterministic) most likely (regular) values. This approach is used in the following analysis. A DEPDF

$$P_h(G,F) = P_0 \exp\left[\left(1 - \frac{G^2}{G_0^2}\right)\exp\left(1 - \frac{F^2}{F_0^2}\right)\right], \quad G \geq G_0, F \geq F_0 \qquad (6.1)$$

of the EVD type (see, e.g., [8]) can be used to characterize the likelihood of a human nonfailure to perform his/her duties, when operating a vehicle [10,11]. Here $P_h(G, F)$ is the probability of nonfailure of the human performance as a function of the off-normal MWL G and outstanding HCF F, P_0 is the probability of nonfailure of the human performance for the specified (normal) MWL $G = G_0$ and the specified (ordinary) HCF $F = F_0$. The specified (most likely, nominal, normal) MWL and HCF can be established by conducting testing and measurements on a flight simulator. It should be emphasized that in this chapter and in this book, the employed DEPDF is always of the EVD "type" but is not necessarily one of the known EVD distributions. It is imperative that the chosen DEPDF for a particular mission or a situation makes physical sense and is able to reflect a particular aerospace mission or a situation and to provide the expected information. The author would like to emphasize also that in a taken approach a suitable and physically meaningful predictive model comes first and is subsequently supported by the relevant statistics, and not the other way around.

The calculated probabilities

$$p = \frac{P_h(G,F)}{P_0} = \exp\left[\left(1 - \frac{G^2}{G_0^2}\right)\exp\left(1 - \frac{F^2}{F_0^2}\right)\right], \quad G \geq G_0, F \geq F_0 \qquad (6.2)$$

(that are, in effect, ratios of the probability of nonfailure in the off-normal conditions to the probability of nonfailure in the normal situation) are shown in Table 6.1.

TABLE 6.1

Calculated Probability Ratios F/F_0 of Human Nonfailure

F/F_0 \ G/G_0	1	2	3	4	5	8	10	∞
1	1	4.979E-2	3.355E-4	3.059E-7	3.775E-11	4.360E-28	1.011E-43	0
2	1	0.8613	0.6715	0.4739	0.3027	0.0434	0.007234	0
3	1	0.9990	0.9973	0.9950	0.9920	0.9791	0.9673	0
4								0
5								0
8				1.0000				0
10								0
∞								1.0000

The following conclusions can be drawn from the table data:

- At normal (specified, most likely) MWL level ($G = G_0$) and/or at an extraordinary (exceptionally) high HCF level ($F \mapsto \infty$) the probability of human nonfailure is close to 100%.

- The probabilities of human nonfailure in off-normal situations are always lower than the probabilities of nonfailure in normal (specified) conditions.

- When the MWL is extraordinarily high, the human will definitely fail, no matter how high his/her HCF is.

- When the HCF is high, even a significant MWL has a small effect on the probability of nonfailure, unless the MWL is exceptionally high. For high HCFs the increase in the MWL has a much smaller effect on the probabilities of failure than for relatively low HCFs.

- The probability of human nonfailure decreases with an increase in the MWL, especially at low MWL levels, and increases with an increase in the HCF, especially at low HCF levels. These intuitively more or less obvious conclusions are quantified by Table 6.1 data.

- These data show also that the increase in the probability ratio above 3.0 ("three is a charm," is it not?) has a minor effect on the probability of nonfailure. This means particularly that the navigator (pilot) does not have to be trained for an unrealistically high MWL (i.e., does not have to be trained by a factor higher than 3.0 compared to a navigator of ordinary capacity [skills, qualification]). In other words, a pilot does not have to be a superman to successfully cope with a high-level MWL, but still has to be trained in such a way that, when there is a need, he/she would be able to cope with a MWL by a factor of 3.0 higher than the normal level, and his/her HCF should be by a factor of 3.0 higher than what is expected of the same person

The "Miracle-on-the-Hudson" Event 137

in ordinary (normal) conditions. Of course, some outstanding individuals (like Captain Sullenberger, for instance) might be characterized by the HCF that corresponds to MWLs somewhat higher than 3.0 (see Table 6.1).

6.5 HCF Needed to Satisfactorily Cope with a High MWL

From (5.2), we obtain

$$\frac{F}{F_0} = \sqrt{1 - \ln\left(\frac{\ln p}{1 - \dfrac{G^2}{G_0^2}}\right)}. \tag{6.3}$$

This relationship is tabulated in Table 6.2. The following conclusion can be drawn from the computed data:

- The HCF level needed to cope with an elevated MWL increases rather slowly with an increase in the probability of nonfailure, especially for high MWL levels, unless this probability is very low (below 0.1) or very high (above 0.9).
- In the region $p = 0.1 \rightarrow 0.9$, the required high HCF level increases with an increase in the MWL level, but this increase is rather moderate, especially for high MWL levels.
- Even for significant MWLs that exceed the normal MWL by orders of magnitude, the level of the HCF does not have to be very much higher than the HCF of a person of ordinary HCF level. When the MWL ratio is as high as 100, the HCF ratio does not have to exceed 4 to assure the probability of nonfailure of as high as 0.999.

TABLE 6.2

Relative HCF F/F_0 versus Relative Probability of Nonfailure and Relative MWL

G/G_0 \ p	E-12	E-3	E-2	0.1	0.5	0.9	0.99	0.9999
5	1.0681	1.4985	1.6282	1.8287	2.1318	2.5354	2.9628	3.6590
10	1.5087	1.9138	2.0169	2.1820	2.4416	2.8010	3.1930	3.8478
100	2.6251	2.8771	2.9467	3.0621	3.2522	3.5300	3.8484	4.4069
1000	3.3907	3.5893	3.6453	3.7392	3.8964	4.1311	4.4063	4.9016
10000	4.0127	4.1819	4.2301	4.3112	4.4483	4.6552	4.9011	5.3508

6.6 Another Approach: Operation Time versus "Available" Landing Time

The above time-independent DEPDF-based approach enables one to compare, on the probabilistic basis, the relative roles of the MWL and HCF in a particular off-normal HITL situation. The role of time (e.g., swiftness in reaction) is accounted for in an indirect fashion, through the HCF level. In the analysis that follows, we assess the likelihood of safe landing by considering the roles of different times directly, by comparing the operation time, which consists of the "subjective" decision-making time and the actual landing time, with the "available" ("objective") landing time (i.e., the time from the moment when an emergency was determined to the moment of actual landing).

Particularly, we address item 10 of Table 6.4 (i.e., the ability of the pilot to anticipate and to make a substantiated and valid decision in a short period of time: "We are going to be in the Hudson"). It is assumed, for the sake of simplicity, that both the decision-making and the landing times could be approximated by the Rayleigh's law, while the available time, considering, in the case of the Miracle-on-the-Hudson flight, the glider conditions of the aircraft, follows the normal law with a high ratio of the mean value to the standard deviation. Safe landing could be expected if the probability that it occurs during the "available" landing time is sufficiently high. The formalism of such a model is similar to the helicopter-landing-ship formalism [9]. If the (random) "subjective" sum, $T = t + \theta$, of the (random) decision-making time, t, and the (random) time, θ, needed to actually land the aircraft is lower, with a high enough probability, than the (random) "objective" duration, L, of the available time, then safe landing becomes possible. In the analysis that follows, we assume the simplest probability distributions for the random times of interest.

We use the Rayleigh's law,

$$f_t(t) = \frac{t}{t_0^2} \exp\left(-\frac{t^2}{2t_0^2}\right), \quad f_\theta(t) = \frac{\theta}{\theta_0^2} \exp\left(-\frac{\theta^2}{2\theta_0^2}\right), \tag{6.4}$$

as a suitable approximation for the random times t and θ of decision-making and actual landing, and the normal law,

$$f_l(l) = \frac{1}{\sqrt{2\pi}\sigma} \exp\left(-\frac{(l-l_0)^2}{2\sigma^2}\right), \quad \frac{l_0}{\sigma} \geq 4.0, \tag{6.5}$$

as an acceptable approximation for the available time, L. In the formulas (6.4) and (6.5), t_0 and θ_0 are the most likely times of decision-making and landing, respectively (in the case of the Rayleigh law these times coincide with the

The "Miracle-on-the-Hudson" Event

standard deviations of the random variables in question), l_0 is the most likely (mean) value of the available time, and σ is the standard deviation of this time. The ratio $\dfrac{l_0}{\sigma}$ ("safety factor") of the mean value of the available time to its standard deviation should be large enough (say, larger than 4), so that the normal law could be used as an acceptable approximation for a random variable that, in principle, cannot be negative, as it is the case when this variable is time. The probability, P_*, that the sum $T = t + \theta$ of the random variable st and θ exceeds a certain time level, $\hat{T}$, can be found on the basis of the convolution of two random times distributed in accordance with the Rayleigh law as follows:

$$P_* = 1 - \int_0^{\hat{T}} \frac{t}{t_0^2} \exp\left(\frac{t^2}{2t_0^2}\right)\left[1 - \exp\left(-\frac{(T-t)^2}{2\theta_0^2}\right)\right]dt$$

$$= \exp\left(-\frac{\hat{T}^2}{2t_0^2}\right) + \exp\left[-\frac{\hat{T}^2}{2(t_0^2 + \theta_0^2)}\right]\left[\frac{\theta_0^2}{t_0^2 + \theta_0^2}\left[\exp\left(-\frac{t_0^2\hat{T}^2}{2\theta_0^2(t_0^2 + \theta_0^2)}\right)\right.\right.$$

$$\left.\left. - \exp\left(-\frac{\theta_0^2\hat{T}^2}{2t_0^2(t_0^2 + \theta_0^2)}\right)\right]\right] + \sqrt{\frac{\pi}{2}}\frac{\hat{T}t_0\theta_0}{(t_0^2 + \theta_0^2)^{3/2}}\exp\left[-\frac{\hat{T}^2}{2(t_0^2 + \theta_0^2)}\right]$$

$$\times\left[\Phi\left(\frac{t_0\hat{T}}{\theta_0\sqrt{2(t_0^2 + \theta_0^2)}}\right) + \Phi\left(\frac{\theta_0\hat{T}}{t_0\sqrt{2(t_0^2 + \theta_0^2)}}\right)\right], \tag{6.6}$$

where

$$\Phi(x) = \frac{2}{\sqrt{\pi}}\int_0^x e^{-z^2}\,dz \tag{6.7}$$

is the Laplace function (probability integral). When the most likely duration of landing, θ_0, is very small compared to the most likely decision-making time, t_0, the expression (6.6) yields

$$P_* = \exp\left(-\frac{\hat{T}^2}{2t_0^2}\right), \tag{6.8}$$

that is, the probability that the total time of operation exceeds a certain time duration, $\hat{T}$, depends only on the most likely decision-making time, t_0. From (6.8), we obtain

$$\frac{t_0}{\hat{T}} = \frac{1}{\sqrt{-2\ln P_*}}. \tag{6.9}$$

If the acceptable probability, P_*, of exceeding the time, $\hat{T}$ (e.g., the available time, if this time is treated as a nonrandom variable of the level $\hat{T}$), is, say, $P = 10^{-4} = 0.01\%$, then the time of making the decision should not exceed $0.2330 = 23.3\%$ of the time, $\hat{T}$ (expected available time), otherwise the requirement $P \leq 10^{-4} = 0.01\%$ will be compromised. If the available time is, say, 2 min, then the decision-making time should not exceed 28 s, which is in agreement with Captain Sullenberger's actual decision-making time. Similarly, when the most likely time, t_0, of decision-making is very small compared to the most likely time, θ_0, of actual landing, the formula (6.6) yields

$$P_* = \exp\left(-\frac{\hat{T}^2}{2\theta_0^2}\right), \tag{6.10}$$

that is, the probability of exceeding a certain time level, $\hat{T}$, depends only on the most likely time, θ_0, of landing. As follows from the formulas (6.4), the probability that the actual time of decision-making or the time of landing exceed the corresponding most likely times is expressed by the formulas of the types (6.8) and (6.10), and is as high as $P_* = \frac{1}{\sqrt{e}} = 0.6065 = 60.6\%$. In this connection, we would like to mention that the one-parametric Rayleigh law is characterized by a rather large standard deviation and therefore might not be the best approximation for the probability density functions for the decision-making time and the time of landing. A more "powerful" and more flexible two-parametric law, such as, for example, the Weibull law, might be more suitable as an appropriate probability distribution of the random times, t and θ. Its use, however, will make our analysis unnecessarily more complicated. Our goal is not so much to "dot all the i's and cross all the t's," as far as modeling of the role of the human factor in the problem in question is concerned, but rather to demonstrate that the attempt to use probabilistic risk management (PRM) methods to quantify the role of the human factor in avionics safety and similar problems might be quite fruitful. When developing practical guidelines and recommendations, a particular law of the probability distribution should be established based on the actual statistical data, and employment of various goodness-of-fit criteria might be needed in detailed statistical analyses.

When the most likely times t_0 and θ_0 required for making the go-ahead decision and for the actual landing, are equal, the formula (6.6) yields:

$$P_* = P_*\left(\frac{t_0}{\hat{T}}, \frac{\theta_0}{\hat{T}}\right) = \exp\left(-\frac{\hat{T}^2}{2t_0^2}\right)\left[1 + \sqrt{\pi}\frac{\hat{T}}{2t_0}\exp\left(\left(\frac{\hat{T}}{2t_0}\right)^2\right)erf\left(\frac{\hat{T}}{2t_0}\right)\right]. \tag{6.11}$$

For large enough $\dfrac{\hat{T}}{t_0}$ ratios $\left(\dfrac{\hat{T}}{t_0} \geq 3\right)$ of the critical time $\hat{T}$ to the most likely decision-making or landing time, the second term in the brackets becomes

The "Miracle-on-the-Hudson" Event 141

large compared to unity. The calculated probabilities of exceeding a certain time level, $\hat{T}$, based on the formula (6.11), are shown in Table 6.3. In the third row of this table we indicate, for the sake of comparison, the probabilities, $P°$, of exceeding the given time, $\hat{T}$, when only the time t_0 or only the time θ_0 is different from zero (i.e., for the special case that is mostly remote from the case $t_0 = \theta_0$ of equal most likely times). Clearly, the probabilities computed for other possible combinations of the times t_0 and θ_0 could be found between the calculated probabilities P_* and $P°$.

The following conclusions can be drawn from Table 6.3 data:

- The probability that the total time of operation (the time of decision-making and the time of landing) exceeds the given time level $\hat{T}$, thereby leading to a casualty, rapidly increases with an increase in the total time of operation.

- The probability of exceeding the time level $\hat{T}$ is considerably higher, when the most likely times of decision-making and of landing are finite and especially when they are close to each other, in comparison with the situation when one of these times is significantly shorter than the other (i.e., zero or next-to-zero). This is particularly true for short operation times, like in the situation in question: the ratio $P_*/P°$ of the probability P_* of exceeding the time level $\hat{T}$ in the case of $t_0 = \theta_0$ to the probability $P°$ of exceeding this level in the case $t_0 = 0$ or in the case $\theta_0 = 0$ decreases rapidly with an increase in the time of operation. There exists, therefore, a significant incentive for reducing the operation time. The importance of this intuitively obvious fact is quantified by the table data.

- Other useful information that could be drawn from the data of the type shown in Table 6.3 is whether it is possible at all to train a human to react (make a decision) in just a couple of seconds. It took Captain Sullenberger about 30 s to make the right decision, and he is an exceptionally highly qualified pilot, with an outstanding HCF. If a very short-term decision could not be expected, and a low

TABLE 6.3

The Probability P_* That the Operation Time Exceeds a Certain Time Level $\hat{T}$ versus the Ratio $\hat{T}/t_0$ of This Time Level to the Most Likely Time t_0 of Decision-Making for the Case When the Time t_0 and the Most Likely Time of Landing θ_0 Are the Same

$\hat{T}/t_0$	6	5	4	3	2
P_*	6.562E-4	8.553E-3	6.495E-2	1.914E-1	6.837E-1
$P°$	1.523E-8	0.373E-5	0.335E-3	1.111E-2	1.353E-1
$P_*/P°$	4.309E4	2.293E3	1.939E2	1.723E1	5.053

For the sake of comparison, the probability $P°$ of exceeding the time level $\hat{T}$, when either the time t_0 or the time θ_0 is zero, is also indicated.

142 *Human-in-the-Loop*

probability of human failure is still required, then one should decide on a broader involvement of more sophisticated, more powerful, and more expensive equipment and instrumentation to do the job. If pursuing such an effort is decided upon, then probabilistic sensitivity analyses of the type developed in the previous text will be needed to determine the most promising ways to go.

It is advisable, of course, that the analytical predictions are confirmed by computer-aided simulations and verified by highly focused and highly cost-effective FOAT conducted on flight simulators. Since the "available" time L is assumed to be a random normally distributed variable, the probability that this time is found below a certain level $\hat{L}$ is

$$P_l = P_l\left(\frac{\sigma}{\hat{L}}, \frac{l_0}{\hat{L}}\right) = \int_{-\infty}^{\hat{L}} f_l(l)dl = \frac{1}{2}\left[1 + \Phi\left(\frac{\hat{L}-l_0}{\sqrt{2}\sigma}\right)\right] = \left[1 + \Phi\left(\frac{1-\frac{l_0}{\hat{L}}}{\sqrt{2}\frac{\sigma}{\hat{L}}}\right)\right]. \quad (6.12)$$

The probability that the available time is exceeded can be determined by equating the times $\hat{T} = \hat{L} = T$ and computing the product

$$P_A = P_*\left(\frac{t_0}{T}, \frac{\theta_0}{T}\right)P_l\left(\frac{\sigma}{T}, \frac{l_0}{T}\right) \quad (6.13)$$

of the probability, $P_*\left(\frac{t_0}{T}, \frac{\theta_0}{T}\right)$, that the time of operation exceeds a certain level, T, and the probability, $P_l\left(\frac{\sigma}{T}, \frac{l_0}{T}\right)$, that the available time is shorter than the time T. The formula (6.13) considers the roles of the most likely available time, the human factor, t_0 (the most likely time required for the pilot to make his/her go-ahead decision), and the most likely time, θ_0, of actual landing (which characterizes both the qualification and skills of the pilot and the qualities/behavior of the flying machine) on the probability of safe landing. Carrying out detailed computations based on the formulas (6.6), (6.12), and (6.13) is, however, beyond the scope of the present chapter.

6.7 Miracle on the Hudson: Incident

US Airways Flight 1549 was a domestic passenger flight from LaGuardia Airport (LGA) in New York City to Charlotte/Douglas International Airport,

The "Miracle-on-the-Hudson" Event

Charlotte, North Carolina. On January 15, 2009, the Airbus A320-214 flying this route struck a flock of Canadian geese during its initial climb out, lost engine power, and ditched in the Hudson River off midtown Manhattan. Since all the 155 occupants survived and safely evacuated the airliner, the incident became known as the "Miracle on the Hudson" [13,14]. The bird strike occurred just northeast of the George Washington Bridge (GWB) about 3 min into the flight and resulted in an immediate and complete loss of thrust from both engines. When the crew determined that they would be unable to reliably reach any airfield, they turned southbound and glided over the Hudson, finally ditching the airliner near the USS Intrepid museum about 3 min after losing power.

The crew was later awarded the Master's Medal of the Guild of Air Pilots and Air Navigators for successful "emergency ditching and evacuation, with the loss of no lives… a heroic and unique aviation achievement…the most successful ditching in aviation history." The pilot in command was 57-year-old Captain Chesley B. "Sully" Sullenberger, a former fighter pilot who had been an airline pilot since leaving the United States Air Force in 1980. He is also a safety expert and a glider pilot. The first officer was Jeffrey B. Skiles, 49. The flight attendants were Donna Dent, Doreen Welsh, and Sheila Dail. The aircraft was powered by two GE Aviation/Snecma-designed CFM56-5B4/P turbofan engines manufactured in France and the United States. One of 74 A320s then in service in the US Airways fleet, it was built by Airbus with final assembly at its facility at Aéroport de Toulouse-Blagnac in France in June 1999 and delivered to the carrier on August 2, 1999. The Airbus is a digital fly-by-wire aircraft: the flight control surfaces are moved by electrical and hydraulic actuators controlled by a digital computer. The computer interprets pilot commands via input from a side-stick, making adjustments on its own to keep the plane stable and on course. This is particularly useful after engine failure by allowing the pilots to concentrate on engine restart and landing planning. The mechanical energy of the two engines is the primary source of electrical power and hydraulic pressure for the aircraft flight control systems. The aircraft also has an auxiliary power unit, which can provide backup electrical power for the aircraft, including its electrically powered hydraulic pumps; and a ram air turbine (RAT), a type of wind turbine that can be deployed into the airstream to provide backup hydraulic pressure and electrical power at certain speeds.

According to the National Transportation Safety Board (NTSB) [14], both the APU and the RAT were operating as the plane descended into the Hudson, although it was not clear whether the RAT had been deployed manually or automatically. The Airbus A320 has a "ditching" button that closes valves and openings underneath the aircraft, including the outflow valve, the air inlet for the emergency RAT, the avionics inlet, the extract valve, and the flow control valve. It is meant to slow flooding in a water landing. The flight crew did not activate the "ditch switch" during the incident. Sullenberger later noted that it probably would not have been effective anyway, since the force of the water impact tore holes in the plane's fuselage much larger than the openings sealed by the switch.

First Officer Skiles was at the controls of the flight when it took off at 3:25 pm, and was the first to notice a formation of birds approaching the aircraft about 2 min later, while passing through an altitude of about 2700 feet (820 m) on the initial climb out to 15,000 feet (4600 m). According to flight data recorder data, the bird encounter occurred at 3:27:11, when the airplane was at an altitude of 2818 feet (856 m) above ground level and at a distance of about 4.5 miles north–northwest of the approach end of runway 22 at LGA. Subsequently, the airplane's altitude continued to increase while the airspeed decreased, until 3:27:30, when the airplane reached its highest altitude of about 3060 feet (930 m), at an airspeed of about 185 knots calibrated airspeed. The altitude then started to decrease as the airspeed started to increase, reaching 210 knots calibrated airspeed at 3:28:10 at an altitude of about 1650 feet (500 m). The windscreen quickly turned dark brown and several loud thuds were heard. Captain Sullenberger took the controls, while Skiles began going through the three-page emergency procedures checklist in an attempt to restart the engines. At 3:27:36 the flight radioed air traffic controllers at New York Terminal Radar Approach Control (TRACON). "Hit birds. We've lost thrust on both engines. We're turning back towards LaGuardia." Responding to the captain's report of a bird strike, controller Patrick Harten, who was working the departure position, told LaGuardia tower to hold all waiting departures on the ground, and gave Flight 1549 a heading to return to LaGuardia. Sullenberger responded that he was unable. Sullenberger asked if they could attempt an emergency landing in New Jersey, mentioning Teterboro Airport in Bergen County as a possibility; air traffic controllers quickly contacted Teterboro and gained permission for a landing on runway 1. However, Sullenberger told controllers that "We can't do it," and that "We're gonna be in the Hudson," making clear his intention to bring the plane down on the Hudson River due to a lack of altitude. Air traffic control at LaGuardia reported seeing the aircraft pass less than 900 feet (270 m) above GWB. About 90 s before touchdown, the captain announced, "Brace for impact," and the flight attendants instructed the passengers how to do so. The plane ended its 6 min flight at 3:31 pm with an unpowered ditching while heading south at about 130 knots (150 mph; 240 km/h) in the middle of the North River section of the Hudson River roughly abeam 50th Street (near the Intrepid Sea-Air-Space Museum) in Manhattan and Port Imperial in Weehawken, New Jersey. Sullenberger said in an interview on CBS television that his training prompted him to choose a ditching location near operating boats so as to maximize the chance of rescue.

After coming to a stop in the river, the plane began drifting southward with the current. NTSB Member Kitty Higgins, the principal spokesperson for the on-scene investigation, said at a press conference the day after the

The "Miracle-on-the-Hudson" Event

accident that it "has to go down [as] the most successful ditching in aviation history... These people knew what they were supposed to do and they did it and as a result, nobody lost their life." The flight crew, particularly Captain Sullenberger, were widely praised for their actions during the incident, notably by New York City Mayor Michael Bloomberg and New York State Governor David Paterson, who opined, "We had a *Miracle on 34th Street*. I believe now we have had a *Miracle on the Hudson*." Outgoing U.S. President George W. Bush said he was "inspired by the skill and heroism of the flight crew," and he also praised the emergency responders and volunteers. Then President-elect Barack Obama said that everyone was proud of Sullenberger's "heroic and graceful job in landing the damaged aircraft," and thanked the A320's crew. The NTSB ran a series of tests using Airbus simulators in France, to see if Flight 1549 could have returned safely to LaGuardia.

The simulation started immediately following the bird strike and "... knowing in advance that they were going to suffer a bird strike and that the engines could not be restarted, four out of four pilots were able to turn the A320 back to LaGuardia and land on Runway 13." When the NTSB later imposed a 30 s delay before they could respond, in recognition that it was not reasonable to expect a pilot to assess the situation and react instantly, all four pilots crashed. On May 4, 2010, the NTSB released a statement that credited the accident outcome to the fact that the aircraft was carrying safety equipment in excess of that mandated for the flight, and excellent cockpit resource management among the flight crew. Contributing factors to the survivability of the accident were good visibility, and fast response from the various ferry operators. Captain Sullenberger's decision to ditch in the Hudson River was validated by the NTSB. On May 28, 2010, the NTSB published its final report into the accident [14]. It determined the cause of the accident to be "the ingestion of large birds into each engine, which resulted in an almost total loss of thrust in both engines."

6.8 Miracle on the Hudson: Flight Segments (Events)

The US AW Flight 1549 events (segments) and durations are summarized (listed) in Table 6.4. It took only 40 s for Captain Sullenberger to make his route change decision and another 2 min to land the aircraft.

6.9 Miracle on the Hudson: Quantitative Aftermath

In this section we intend to demonstrate how the Miracle-on-the-Hudson event could be quantified using the DEPDF-based evaluations.

TABLE 6.4

US AW Flight 1549, January 15, 2009 (Wikipedia)

Flight Segment	Time (EST)	Duration (s)	Altitude (m)	Speed (km/h)	Event
1	3:25:00 pm	60.00 (16.6667)	0	279.6	Aircraft took off LGA and started climbing out. First Officer Skiles runs the aircraft.
2	3:26:00 pm	71.00 (19.7222)	820	—	Skiles noticed a flock of birds.
3	3:27:11 pm	19.00 (5.2778)	856	322.2	Bird strike (North-East of GWB, NYC).
4	3:27:30 pm	6.00 (1.6667)	930	342.6	Highest altitude reached.
5	3:27:36 pm	24.00 (6.6667)	—	359.3	Radioed TRACON traffic controllers: "Hit birds. Lost thrust on both engines. Turning back towards LGA."
6	3:28:00 pm	10.00 (2.7778)	609	374.1	Complete loss of thrust (engine power).
7	3:28:10 pm	30.00 (8.3333)	500	388.9	Sullenberger takes over control.
8	3:28:40 pm	20.00 (5.5555)	500	388.9	Sullenberger makes route change decision and turns southbound.
9	3:29:00 pm	10.00 (2.7778)	396	353.7	Started gliding over Hudson River.
10	3:29:10 pm	90.00 (25.0000)	—	—	"Brace for impact" command.
11	3:30:40 pm	20.00 (5.5555)	0	240	Touch down (ditching) Hudson River.
12	3:31:00 pm	—	0	0	Full stop, start drifting.

As evident from the computed data, the probability of human nonfailure in off-normal flight conditions is still relatively high, provided that the HCF is significantly higher than that of a pilot of normal skills in the profession and that the MWL is not extraordinarily (perhaps, unrealistically) high. So, the actual "miraculous" event was due to the fact that a person of extraordinary abilities (measured by the level of the HCF) turned out to be in the driving chair at the critical moment. Other favorable aspects of the situation were high HCF of the crew, good weather, and the landing site, perhaps the most favorable one could imagine.

As long as this miracle did happen, everything else was not really a miracle. Captain Sullenberger knew when to take over control of the aircraft, when to abandon his communications with the (generally speaking, excellent) air traffic control (ATC) and to use his outstanding background and skills to land (ditch) the plane: "I was sure I could do it…the entire life up to this moment was a preparation for this moment…I am not just a pilot of that flight. I am also a pilot who has flown for 43 years…" Such a "miracle" does not happen often, of course, and is perhaps outside any indicative statistics.

6.10 Captain Sullenberger

Sullenberger was born to a dentist father—a descendant of Swiss immigrants named Sollenberger—and an elementary schoolteacher mother. He has one sister, Mary Wilson. The street on which he grew up in Denison, Texas, was named after his mother's family, the Hannas. According to his sister, Sullenberger built model planes and aircraft carriers during his childhood, and might have become interested in flying after hearing stories about his father's service in the United States Navy. He went to school in Denison, and was consistently in the 99th percentile in every academic category. At the age of 12, his IQ was deemed high enough to join Mensa International. He also gained a pilot's license at 14. In high school he was the president of the Latin club, a first chair flute, and an honor student. His high school friends have said that Sullenberger developed a passion for flying from watching jets based out of Perrin Air Force Base. Sullenberger's hypothetical HCF is computed in Table 6.5. The calculations of the probability of the human nonfailure are carried out using formula (6.2) and are shown in Table 6.6. We did not try to anticipate and quantify a particular (most likely) MWL level, but rather assumed different MWL deviations from the most likely level.

A more detailed MWL analysis can be done using flight simulation FOAT data. The computed data indicate that, as long as the HCF is high, even significant relative MWL levels, up to 50 or even higher, still result in a rather high probability of the human nonfailure. Captain Sullenberger's HCF is/was extraordinarily, exceptionally high. This was due to his age, old enough to be an experienced performer and young enough to operate effectively in a cool demeanor under pressure and possess other qualities of a relatively young human. Of course, there are other highly qualified pilots who might be able to do what Captain Sullenberger did, but this guess is even more hypothetical than our hypothetical estimates of Captain Sullenberger's HCF qualities.

6.11 Flight Attendant's Hypothetical HCF Estimate

The HCF of a flight attendant is assessed in Table 6.7, and the probabilities of his/her nonfailure are shown in Table 6.8.

The qualities expected from a flight attendant are, of course, quite different than those of a pilot. As evident from the obtained data, the probability of the human nonfailure of the Airbus A-320 flight attendants is rather high up until the MWL ratio of 10 or even slightly higher. Although we do not try to evaluate the first officer's (Skiles) HCF, we assume that his HCF is also high, although this did not manifest itself during the event.

TABLE 6.5

Sullenberger's Hypothetical HCF

Number	Relevant Qualities	Relative HCF Rating $\left(\dfrac{F^*}{F_0}\right)$	Comments
1	Psychological suitability for the given task	3.2	1. 57-year-old former fighter pilot who had been a commercial airline pilot since leaving the United States Air Force in 1980. He is also a *safety* expert and a *glider* pilot [5.7]. See also Appendix B. "I was sure I could do it." "The entire life up to this moment was a preparation for this moment." "I am not just a pilot of that flight. I am also a pilot who has flown for 43 years...."
2	Professional qualifications and experience	3.9	
3	Level, quality, and timeliness of past and recent training	2.0	
4	Mature (realistic) and independent thinking	3.2	
5	Performance sustainability (predictability, consistency)	3.2	
6	Ability to concentrate and act in cold blood ("cool demeanor") in hazardous and even in life-threatening situations	3.3	
7	Ability to anticipate ("expecting the unexpected")	3.2	2. Probability of human nonfailure in normal flight conditions is assumed to be 100%
8	Ability to operate effectively under pressure	3.4	3. The formula $$p = \exp\left(1 - \frac{G^2}{G_0^2}\right)$$ would have to be used to evaluate the probability of nonfailure in the case of a pilot of ordinary skills. The computed numbers are shown in parentheses. The computed numbers show that such a pilot would definitely fail in the off-normal situation in question.
9	Self-control in hazardous situations	3.2	
10	Ability to make a substantiated decision in a short period of time ("we are going to be in the Hudson")	2.8	
	Average FOM	3.14	

$(^*)$ This is just an example that shows that the approach makes physical sense. Actual numbers should be obtained using FOAT on a simulator and confirmed by an independent approach, such as, say, Delphi method: http://en.wikipedia.org/wiki/Delphi_method [12].

TABLE 6.6

Computed Probabilities of Human Nonfailure (Captain Sullenberger)

G/G_0	5	10	50	100	150
p	0.9966	0.9860	0.7013	0.2413	0.0410

It has been shown elsewhere [10] that it is expected that both pilots have high and, to an extent possible, equal qualifications and skills for a high probability of a mission success, if, for one reason or another, the entire MWL is taken by one of the pilots.

The "Miracle-on-the-Hudson" Event

TABLE 6.7

Flight Attendant's HCF

Number	Relevant Qualities	Relative HCF $\left(\dfrac{F^*}{F_0}\right)$
1	Psychological suitability for the task	2.5
2	Professional qualifications and experience	2.5
3	Level, quality, and timeliness of past and recent training	2.5
4	Team-player attitude	3.0
5	Performance sustainability (consistency)	3.0
6	Ability to perform in cold blood in hazardous and even in life-threatening situations	3.0
7	Ability and willingness to follow orders	3.0
8	Ability to operate effectively under pressure	3.4
	Average FOM	2.8625

(*)It is just an example. Actual numbers should be obtained using FOAT on a simulator and confirmed by an independent method, such as, say, Delphi method: http://en.wikipedia.org/wiki/Delphi_method [12].

TABLE 6.8

Estimated Probabilities of Nonfailure for a Flight Attendant

G/G_0	5	10	50	100	150
p	0.9821	0.9283	0.1530	5.47E-4	4.57E-8

In this connection, we would like to mention that, even regardless of the qualification, it is widely accepted in the avionic and maritime practice that it is the captain, not the first officer (first mate) who gets in control of dangerous situations, especially life-threatening ones. It did not happen, however, in the case of the Swissair UN-shuttle last flight addressed in the next section.

6.12 UN-Shuttle Flight: Crash

For the sake of comparison of the successful Miracle-on-the-Hudson case with an emergency situation that ended up in a crash, we have chosen the infamous Swissair September 2, 1998, Flight 111, when a highly trained crew made several bad decisions under considerable time pressure [15] that was, however, not as severe as in the Miracle-on-the-Hudson case.

Swissair Flight 111 was a McDonnell Douglas MD-11 on a scheduled airline flight from John F. Kennedy (JFK) International Airport in New York City to Cointrin International Airport in Geneva, Switzerland.

On Wednesday, September 2, 1998, the aircraft crashed into the Atlantic Ocean southwest of Halifax International Airport at the entrance to St. Margaret's Bay, Nova Scotia. The crash site was just 8 km (5 nm) from shore. All 229 people on board died—the highest death toll of any aviation accident involving a McDonnell Douglas MD-11.

Swissair Flight 111 was known as the "UN-shuttle" due to its popularity with United Nations officials; the flight often carried business executives, scientists, and researchers. The initial search and rescue response, crash recovery operation, and resulting investigation by the Government of Canada took over 4 years. The Transportation Safety Board of Canada's official report stated that flammable material used in the aircraft's structure allowed a fire to spread beyond the control of the crew, resulting in the loss of control and crash of the aircraft. An MD-11 has a standard flight crew consisting of a captain and a first officer, and a cabin crew made up of a maître-de-cabine (M/C, purser) supervising the work of 11 flight attendants. All personnel on board Swissair Flight 111 were qualified, certified, and trained in accordance with Swiss regulations, under the Joint Aviation Authorities. The flight details are shown in Table 6.9.

The flight took off from New York's JFK Airport at 20:18 Eastern Standard Time (EST). Beginning at 20:33 EST and lasting until 20:47, the aircraft experienced an unexplained 13 min radio blackout. The cause of the blackout, or if it was related to the crash, is unknown. At 22:10 Atlantic Time (21:10 EST), cruising at FL330 (approximately 33,000 feet or 10,100 m), Captain Urs Zimmermann and First Officer Stephan Loew detected an odor in the cockpit and determined it to be smoke from the air-conditioning system, a situation easily remedied by closing the air-conditioning vent, which a flight attendant did on Zimmermann's request. Four minutes later, the odor returned and now smoke was visible, and the pilots began to consider diverting to a nearby airport for the purpose of a quick landing. At 22:14 AT (21:14 EST) the flight crew made a radio call to ATC at Moncton (which handles trans-Atlantic air traffic approaching or departing North American air space), indicating that there was an urgent problem with the flight, although not an emergency, which would imply immediate danger to the aircraft. The crew requested a diversion to Boston's Logan International Airport, which was 300 nautical miles (560 km) away. ATC Moncton offered the crew a vector to the closer, 66 nm (104 km) away, Halifax International Airport in Enfield, Nova Scotia, which Loew accepted. The crew then put on their oxygen masks and the aircraft began its descent. Zimmermann put Loew in charge of the descent, while he personally ran through the two Swissair standard checklists for smoke in the cockpit, a process that would take approximately 20 min and become a later source of controversy.

At 22:18 AT (21:18 EST), ATC Moncton handed over traffic control of Swissair 111 to ATC Halifax, since the plane was now going to land in Halifax rather than leave North American air space. At 22:19 AT (21:19 EST) the plane was 30 nautical miles (56 km) away from Halifax International Airport, but Loew

The "Miracle-on-the-Hudson" Event   151

TABLE 6.9

Swissair Flight 111, September 2, 1998 (Wikipedia)

Flight Segment	Time (EST)	Event
1	20:18:00	Aircraft took off JFK airport. First Officer Stephan Loew runs the aircraft.
2	20:33–20:47	Radio blackout.
3	21:10	Captain Urs Zimmermann and First Officer Stephan Loew detected an odor in the cockpit and determined it to be smoke from the air-conditioning system, a situation easily remedied by closing the air-conditioning vent, which a flight attendant did on Zimmermann's request.
4	21:14	Odor returned and smoke became visible. The crew called ATC Moncton indicating an urgent, but not an emergency, problem, and requested a diversion to Boston's Logan Airport, which was 300 nm (560 km) away. ATC Moncton offered a vector to the closer Halifax Airport in Enfield, Nova Scotia, 66 nm (104 km) away, which Loew accepted.
5	21:14–21:34	The crew put on oxygen masks and the aircraft began to descent. Zimmermann put Loew in charge of the descent, while he ran through the Swissair checklists for smoke in the cockpit, a process that later became a source of controversy.
6	21:18	ATC Moncton handed over traffic control of Swissair 111 to ATC Halifax.
7	21:19	The plane was 30 nm (56 km) away from Halifax Airport, but Loew requested more time to descend the plane from its altitude of 6400 m.
8	21:20	Loew informed ATC Halifax that he needed to dump fuel. ATC Halifax said later it was a surprise, because the request came so late. Dumping fuel was a fairly standard procedure early on in nearly any "heavy" aircraft urgent landing scenario. Subsequently, ATC Halifax diverted aircraft toward St. Margaret's Bay, where they could more safely dump fuel, but still be only around 30 nm (56 km) from Halifax.
9	21:24:28	In accordance with the Swissair "In case of smoke of unknown origin" checklist, the crew shut off the power supply in the cabin. This caused the recirculating fans to shut off. This caused a vacuum, which induced the fire to spread back into the cockpit. This also caused the autopilot to shut down. Loew informed ATC Halifax that "we now must fly manually."
10	21:24:45	Loew informed ATC Halifax that "Swissair 111 is declaring emergency."
11	21:24:46	Loew repeated the emergency declaration 1 s later, and over the next 10 s stated that they had descended to "between 12,000 and 5,000 feet" and once more declared an emergency.
12	21:25:40	The flight data recorder stopped recording, followed 1 s later by the cockpit voice recorder.
13	21:25:50–21:26:04	The doomed plane briefly showed up again on radar screens. Its last recorded altitude was 9700 feet. Shortly after the first emergency declaration, the captain could be heard leaving his seat to fight the fire, which was now spreading to the rear of the cockpit.

requested more time to descend the plane from its altitude of 21,000 feet (6400 m). At 22:20 AT (21:20 EST), Loew informed ATC Halifax that he needed to dump fuel, which ATC Halifax controllers would say later, was a surprise considering that the request came so late; dumping fuel is a fairly standard procedure early on in nearly any "heavy" aircraft urgent landing scenario. ATC Halifax subsequently diverted Swissair 111 toward St. Margaret's Bay, where they could more safely dump fuel, but still be only around 30 nautical miles (56 km) from Halifax. In accordance with the Swissair checklist entitled "In case of smoke of unknown origin," the crew shut off the power supply in the cabin, which caused the recirculating fans to shut off. This caused a vacuum, which induced the fire to spread back into the cockpit. This also caused the autopilot to shut down; at 22:24:28 AT (21:24:28 EST), Loew informed ATC Halifax that "we now must fly manually." Seventeen seconds later, at 22:24:45 AT (21:24:45 EST), Loew informed ATC Halifax that "Swissair 111 heavy is declaring emergency," repeated the emergency declaration 1 s later, and over the next 10 s stated that they had descended to "between 12,000 and 5,000 feet" and once more declared an emergency. The flight data recorder stopped recording at 22:25:40 AT (21:25:40 EST), followed 1 s later by the cockpit voice recorder. The doomed plane briefly showed up again on radar screens from 22:25:50 AT (21:25:50 EST) until 22:26:04 AT (21:26:04 EST). Its last recorded altitude was 9700 feet. Shortly after the first emergency declaration, the captain could be heard leaving his seat to fight the fire, which was now spreading to the rear of the cockpit. The Swissair volume of checklists was later found fused together, as if someone had been trying to use them to fan back flames. The captain did not return to his seat, and whether he was killed from the fire or asphyxiated by the smoke is not known. However, physical evidence provides a strong indication that First Officer Loew may have survived the inferno only to die in the eventual crash; instruments show that Loew continued trying to fly the now-crippled aircraft, and gauges later indicated that he shut down engine two approximately 1 min before impact, implying he was still alive and at the controls until the aircraft struck the ocean at 22:31 AT (21:31 EST). The aircraft disintegrated on impact, killing all on board instantly. The search and rescue operation was launched immediately by Joint Rescue Coordination Centre Halifax which tasked the Canadian Forces Air Command, Maritime Command and Land Force Command, as well as Canadian Coast Guard and Canadian Coast Guard Auxiliary resources. The first rescue resources to approach the crash site were Canadian Coast Guard Auxiliary volunteer units—mostly privately owned fishing boats—sailing from Peggy's Cove, Bayswater, and other harbors on St. Margaret's Bay and the Aspotogan Peninsula. They were soon joined by the dedicated Canadian Coast Guard SAR vessel CCGS *Sambro* and CH-113 Labrador SAR helicopters flown by 413 Squadron from CFB Greenwood. The investigation identified 11 causes and contributing factors of the crash in its final report. The first and most important was the following: "Aircraft certification standards for material flammability were inadequate in that they allowed the use of materials

The "Miracle-on-the-Hudson" Event

that could be ignited and sustain or propagate fire. Consequently, flammable material propagated a fire that started above the ceiling on the right side of the cockpit near the cockpit rear wall. The fire spread and intensified rapidly to the extent that it degraded aircraft systems and the cockpit environment, and ultimately led to the loss of control of the aircraft." Arcing from wiring of the in-flight entertainment system network did not trip the circuit breakers.

While suggestive, the investigation was unable to confirm if this arc was the "lead event" that ignited the flammable covering on metalized poly-ethylene terephthalate insulation blankets that quickly spread across other flammable materials. The crew did not recognize that a fire had started and were not warned by instruments. Once they became aware of the fire, the uncertainty of the problem made it difficult to address. The rapid spread of the fire led to aircraft. Because he had no light by which to see his controls after the displays failed, the pilot was forced to steer the plane blindly; intentionally or not, the plane swerved off course and headed back out into the Atlantic.

Recovered fragments of the plane show that the heat inside the cockpit became so great that the ceiling started to melt. The recovered standby attitude indicator and airspeed indicator showed that the aircraft struck the water at 300 knots (560 km/h, 348 mph) in a 20° nose down and 110° bank turn, or almost upside down. Less than a second after impact the plane would have been totally crushed, killing all aboard almost instantly. The Transportation Safety Board concluded that even if the crew had been aware of the nature of the problem, the rate at which the fire spread would have precluded a safe landing at Halifax even if an approach had begun as soon as the "pan-pan-pan" was declared. The plane was broken into two million small pieces by the impact, making this process time-consuming and tedious. The investigation became the largest and most expensive transport accident investigation in Canadian history.

6.13 Swissair Flight 111: Segments (Events) and Crew Errors

The Swissair Flight 111 events (segments) and durations are summarized in Table 6.9. The following more or less obvious errors were made by the crew:

- At 21:14 EST they used poor judgment and underestimated the danger by indicating to the ATC Moncton that the returned odor and visible smoke in the cockpit was an urgency, but not an emergency problem. They requested a diversion to the 300 nm (560 km) away Boston Logan Airport, and not to the closest 66 nm (104 km) away Halifax Airport.

TABLE 6.10

Hypothetical HCF for the Flight 111 Pilot

Number	Relevant Qualities	HCF
1	Psychological suitability for the given task	3.0
2	Professional qualifications and experience	3.0
3	Level, quality, and timeliness of past and recent training	2.0
4	Mature (realistic) and independent thinking	1.0
5	Performance sustainability (consistency)	2.0
6	Ability to concentrate and to act in cold blood in hazardous situations	1.5
7	Ability to anticipate ("expecting the unexpected")	1.2
8	Ability to operate effectively under pressure	1.5
9	Self-control in hazardous situations	2.0
10	Ability to make a substantiated decision in a short period of time	1.2
	Average FOM	1.84

(*)It is just an example. Actual numbers should be obtained using FOAT on a simulator and confirmed by an independent method, such as, say, Delphi method: http://en.wikipedia.org/wiki/Delphi_method [12].

- Captain Zimmermann put First Officer Loew in charge of the descent and spent time running through the Swissair checklist for smoke in the cockpit.
- At 21:19 EST Loew requested more time to descend the plane from its altitude of 6400 m, although the plane was only 30 nm (56 km) away from Halifax Airport.
- At 21:20 EST Loew informed ATC Halifax that he needed to dump fuel. As ATC Halifax indicated later, it was a surprise, because the request came too late. In addition, it was doubtful that such a measure was needed at all.
- At 21:24:28 the crew shut off the power supply in the cabin. That caused the recirculating fans to shut off and caused a vacuum, which induced the fire to spread back into the cockpit. This also caused the autopilot to shut down, and Loew had to "fly manually." In about a minute or so the plane crashed.

These errors are reflected in Table 6.10 score sheet and resulted in a rather low HCF and low probability of the assessed human nonfailure.

6.14 Flight 111 Pilot's Hypothetical HCF

Flight 111 pilot's HCF and the probability of human nonfailure are summarized in Table 6.10. The criteria used are the same as in Table 6.5. The

TABLE 6.11

Computed Probabilities of Human Nonfailure (Swissair Pilot)

G/G_0	5	10	50	100
p	0.1098	1.1945E-4	0	0

probabilities of human nonfailure are shown in Table 6.11. The computed probability of nonfailure is very low even at not very high MWL levels. Although the crew's qualification seems to be adequate, qualities 4, 6, 7, 8, and 10, which were particularly critical in the situation in question, turned out to be extremely low. No wonder that it led to a crash.

6.15 Other Reported Water Landings of Passenger Airplanes

- On July 11, 2011, Angara Airlines Flight 5007 (an Antonov An-24) ditched in the Ob River near Strezhevoy, Russia, after an engine fire. Upon water contact the tail separated and the burnt port engine became detached from its mounts. Otherwise the plane remained intact, but was written off. Out of 37 people on board, including 4 crew and 33 passengers, 7 passengers died. Of the survivors, at least 20 were hospitalized with various injuries.

- On June 6, 2011, a Solenta Aviation Antonov An-26 freighter flying for DHL Aviation ditched in the Atlantic Ocean near Libreville, Gabon. Three crew and one passenger were rescued with minor injuries.

- On October 22, 2009, a Divi Divi Air Britten-Norman Islander operating Divi Divi Air Flight 014 ditched in off the coast of Bonaire after its starboard engine failed. The pilot reported that the aircraft was losing 200 feet per minute after choosing to fly to an airport. All nine passengers survived, but the captain was knocked unconscious and although some passengers attempted to free him, he drowned and was pulled down with the aircraft.

- On August 6, 2005, Tuninter Flight 1153 (an ATR 72) ditched off the Sicilian coast after running out of fuel. Of 39 aboard, 23 survived with injuries. The plane's wreck was found in three pieces.

- On January 16, 2002, Garuda Indonesia Flight 421 (a Boeing 737) successfully ditched into the Bengawan Solo River near Yogyakarta, Java Island, after experiencing a twin engine flameout during heavy precipitation and hail. The pilots tried to restart the engines several times before making the decision to ditch the aircraft. Photographs taken shortly after evacuation show that the plane came to rest in knee-deep water. Of the 60 occupants, one flight attendant was killed.

- On November 23, 1996, Ethiopian Airlines Flight 961 (a Boeing 767-260ER), ditched in the Indian Ocean near Comoros after being hijacked and running out of fuel, killing 125 of the 175 passengers and crew on board. Unable to operate flaps, it impacted at high speed, dragging its left wing tip before tumbling and breaking into three pieces. The panicking hijackers were fighting the pilots for the control of the plane at the time of the impact, which caused the plane to roll just before hitting the water, and the subsequent wing tip hitting the water and breakup are a result of this struggle in the cockpit. Some passengers were killed on impact or trapped in the cabin when they inflated their life vests before exiting. Most of the survivors were found hanging onto a section of the fuselage that remained floating.

- On May 2, 1970, ALM Flight 980 (a McDonnell Douglas DC-9-33CF), ditched in mile-deep water after running out of fuel during multiple attempts to land at Princess Juliana International Airport on the island of Saint Maarten in the Netherlands Antilles under low-visibility weather. Insufficient warning to the cabin resulted in several passengers and crew still either standing or with unfastened seat belts as the aircraft struck the water. Of 63 occupants, 40 survivors were recovered by U.S. military helicopters.

- On August 21, 1963, an Aeroflot Tupolev Tu-124 ditched into the Neva River in Leningrad (now St. Petersburg) after running out of fuel. The aircraft floated and was towed to shore by a tugboat, which it had nearly hit as it came down on the water. The tug rushed to the floating aircraft and pulled it with its passengers near to the shore, where the passengers disembarked onto the tug; all 52 on board escaped without injuries.

- On September 23, 1962, *Flying Tiger Line* Flight 923, a Lockheed 1049H-82 Super Constellation N6923C, passenger aircraft, on a military (MATS) charter flight, with a crew of 8 and 68 U.S. civilian and military (paratrooper) passengers, ditched in the North Atlantic about 500 miles west of Shannon, Ireland, after losing three engines on a flight from Gander, Newfoundland, to Frankfurt, West Germany. Forty-five of the passengers and 3 crew were rescued, with 23 passengers and 5 crew members being lost in the storm-swept seas. All occupants successfully evacuated the airplane. Those who were lost succumbed in the rough seas.

The "Miracle-on-the-Hudson" Event

- In October 1956, Pan Am Flight 6 (a Boeing 377) ditched northeast of Hawaii, after losing two of its four engines. The aircraft was able to circle around USCGC *Pontchartrain* until daybreak, when it ditched; all 31 on board survived.

- In April 1956, Northwest Orient Airlines Flight 2 (also a Boeing 377) ditched into Puget Sound after what was later decided to be caused by failure of the crew to close the cowl flaps on the plane's engines. All aboard escaped the aircraft after a textbook landing, but four passengers and one flight attendant succumbed either to drowning or to hypothermia before being rescued.

- On March 26, 1955, Pan Am Flight 845/26 ditched 35 miles from the Oregon coast after an engine tore loose. Despite the tail section breaking off during the impact, the aircraft floated for 20 min before sinking. Survivors were rescued after a further 90 min in the water.

- On June 19, 1954, Swissair Convair CV-240 HB-IRW ditched into the English Channel because of fuel starvation, which was attributed to pilot error. All three crew and five passengers survived the ditching and could escape the plane. However, three of the passengers could not swim and eventually drowned, because there were no life jackets on board, which was not prescribed at the time.

- On August 3, 1953, Air France Flight 152, a Lockheed L-749A Constellation, ditched 6 miles from Fetiye Point, Turkey 1.5 miles offshore into the Mediterranean Sea on a flight between Rome, Italy, and Beirut, Lebanon. The propeller had failed due to blade fracture. Due to violent vibrations, engine number three broke away and control of engine number four was lost. The crew of eight and all but four of the 34 passengers were rescued.

- On April 16, 1952, the de Havilland Australia DHA-3 Drover VH-DHA operated by the Australian Department of Civil Aviation with three occupants was ditched in the Bismarck Sea between Wewak and Manus Island. The port propeller failed, a propeller blade penetrated the fuselage, and the single pilot was rendered unconscious; the ditching was performed by a passenger.

- On April 11, 1952, Pan Am Flight 526A ditched 11.3 miles northwest of Puerto Rico due to engine failure after takeoff. Many survived the initial ditching, but panicking passengers refused to leave the sinking wreck and drowned. Fifty-two passengers were killed, 17 passengers and crew members were rescued by the U.S. Coast Guard. After this accident it was recommended to implement preflight safety demonstrations for overwater flights.

6.16 Conclusions

The following conclusions can be drawn from the performed analyses:

- Application of a quantitative PRM approach should complement, whenever feasible and possible, the existing vehicular psychology practices that are, as a rule, qualitative assessments of the role of the human factor when addressing the likelihood of success and safety of various vehicular missions and situations.

- It has been the high HCF of the aircraft crew and especially of Captain Sullenberger that made a reality what seemed to be a "miracle." The carried-out PRM-based analysis enables to quantify this fact. In effect, it has been a "miracle" that an outstanding individual like Captain Sullenberger turned out to be in control at the time of the incident and that the weather and the landing location were highly favorable. As long as this took place, nothing else could be considered as a "miracle": the likelihood of safe landing with an individual like Captain Sullenberger in the cockpit was very high.

- The taken PRM-based approach, after the trustworthy input information is obtained using FOAT on a simulator and confirmed by an independent approach, such as, say, Delphi method, is applicable to many other HITL situations, well beyond the situation in question and perhaps even beyond the vehicular domain.

- Although the obtained numbers make physical sense, it is the approach, not the actual numbers, that is, in the author's opinion, the merit of the analysis/message.

References

1. J.T. Reason, *Human Error*, Cambridge University Press, Cambridge, 1990.
2. A.T. Kern, *Controlling Pilot Error: Culture, Environment, and CRM* (Crew Resource Management), McGraw-Hill, New-York, 2001.
3. W.A. O'Neil, "The Human Element in Shipping," Keynote Address. In *Biennial Symposium of the Seafarers International Research Center*, Cardiff, Wales, June 29, 2001.
4. D.C. Foyle and B.L. Hooey, *Human Performance Modeling in Aviation*, CRC Press, Boca Raton, FL, 2008.
5. D. Harris, *Human Performance on the Flight Deck*, Bookpoint Ltd., Ashgate Publishing, Oxon, 2011.
6. E. Hollnagel, *Human Reliability Analysis: Context and Control*, Academic Press, London and San Diego, 1993.

7. J.T. Reason, *Managing the Risks of Organizational Accidents*, Ashgate Publishing Company, Burlington, VT, 1997.
8. E. Suhir, *Applied Probability for Engineers and Scientists*, McGraw-Hill, New York, 1997.
9. E. Suhir, "Helicopter-Landing-Ship: Undercarriage Strength and the Role of the Human Factor." In *ASME OMAE Conference*, June 1–9, Honolulu, Hawaii, 2009; see also *ASME OMAE Journal*, February 2010.
10. E. Suhir and R.H. Mogford, "'Two Men in a Cockpit': Probabilistic Assessment of the Likelihood of a Casualty if One of the Two Navigators Becomes Incapacitated,"*Journal of Aircraft*, vol. 48, No. 4, July–August 2011.
11. E. Suhir, "Human-in-the-Loop": Likelihood of a Vehicular Mission-Success-and-Safety, and the Role of the Human Factor,"Paper ID 1168, 2011 IEEE/AIAA Aerospace Conference, Big Sky, MT, March 5–12, 2011; see also *Journal of Aircraft*, vol. 49, No. 1, 2012.
12. http://en.wikipedia.org/wiki/Delphi_method.
13. http://en.wikipedia.org/wiki/US_Airways_Flight_1549.
14. http://www.ntsb.gov/doclib/reports/2010/AAR1003.pdf.
15. http://en.wikipedia.org/wiki/Swissair_Flight_111.

7

"Two Men in a Cockpit": Likelihood of a Casualty If One of the Pilots Gets Incapacitated

7.1 Summary

Using the *double-exponential probability distribution function* (DEPDF), we consider a situation when two pilots operate an aircraft in an ordinary (normal, routine) fashion that abruptly changes to an extraordinary (off-normal, hazardous) one when one of the pilots becomes, for one reason or another, incapacitated. Such a mishap is referred to as an *accident*. Because of the accident, the other pilot, *pilot-in-charge* (PIC), might have to cope with a higher *mental workload* (MWL). A fatal *casualty* will occur if both pilots become incapable of performing their duties.

Although this circumstance will eventually manifest itself most likely only during landing, we, nonetheless, consider, in order to assess the probability of the potential casualty, an en route situation (i.e., a situation that precedes descending and landing). This probability depends on the capability of the PIC to successfully cope with the increased MWL. We determine the probability of a casualty as a function of the actual MWL level and the level of the *human capacity factor* (HCF). The total flight time and the time after the accident are treated as nonrandom parameters.

We believe that the suggested MWL/HCF model and its generalizations, after appropriate sensitivity analyses are carried out, can be helpful when developing guidelines for personnel training; when choosing the appropriate flight simulation conditions; and/or when there is a need to decide if the existing level of automation and the navigation instrumentation and equipment are adequate to cope with an extraordinary (off-normal) situation. If not, additional and/or more advanced and more expensive instrumentation and equipment should be considered, developed, and installed. Plenty of additional risk analyses and human psychology related efforts will be needed, of course, to make the guidelines based on the suggested probabilistic risk analysis (PRA) model practical.

161

7.2 Introduction

In this analysis we use a DEPDF to quantify the likelihood of a human failure to perform his/her duties, when operating a vehicle. We consider a situation when two pilots operate an aircraft in an ordinary (normal, routine) fashion that abruptly changes, because of a *mishap*, when one of the pilots becomes incapacitated, to an extraordinary (off-normal, hazardous) situation. A fatal *casualty* will occur if both pilots become incapable of performing their duties. Although this circumstance will most likely manifest itself only during landing, which is the phase of flight with the highest MWL, we, nonetheless, address the en route condition (that precedes descending and landing) to determine how the likelihood of the PIC failure to perform his/her normal and off-normal duties might affect the probability of the potential casualty. We assess such a likelihood as a function of the operator's MWL and HCF. Our analysis is, in effect, an attempt to quantify, on the probabilistic basis, using a suitable analytical method, the human's ability (capacity) to cope with an elevated MWL.

The suggested DEPDF considers the duration of the mission (flight, journey, task, operation); the time in operation prior to, and after, the accident (when the MWL suddenly increases); and the relative levels of the MWL and the HCF (i.e., the human's ability to cope with the off-normal MWL). In our approach, the relative levels of the MWL and HCF (these could be steady state or time dependent) determine the likelihood of the mission success (safety). Cognitive overload has been recognized as a significant cause of error in aviation, and therefore, measuring the MWL has become a key method of improving safety. There is an extensive published work in the psychological literature devoted to the measurement of MWL, both in military and in civil aviation (see, e.g., Refs. [1–18]). A pilot's MWL can be measured using subjective ratings or objective measures [1,4]. The subjective ratings could be carried out during simulation tests and can be in the form of periodic inputs to some kind of data collection device that, for example, prompts the pilot to enter a number between 1 and 10 to estimate the MWL every few minutes. Another possible approach is postflight paper questionnaires. There are also some objective measures of MWL, such as, for example, heart rate variability. It is easier to measure the MWL on a flight simulator than in actual flight conditions. In the real world, an analyst would probably be restricted to using postflight subjective (questionnaire) measures, in order not to interfere with the pilot's work. An aircraft pilot faces numerous challenges imposed by the need to control a multivariate lagged system in a heterogeneous multitask environment. The time lags between critical variables require predictions and actions in an uncertain world.

The interrelated concepts of situation awareness and MWL are central to aviation psychology. The major components of situation awareness are spatial awareness, system awareness, and task awareness. Each of these components has real-world implications: spatial awareness for instrument displays, system awareness for keeping the operator informed about actions that have been taken by automated systems, and task awareness for attention and task management. Task management is directly related to the level of the MWL, as the competing "demands" of the tasks for attention might exceed the operator's resources—his/her "capacity" to adequately cope with the "demands" imposed by the MWL.

In modern military aircraft, complexity of information, combined with time stress, creates difficulties for the pilot under combat conditions, and the first step to mitigate this problem is to measure and manage the MWL. Although there is no universally accepted definition of the MWL and how it should/could be evaluated, there is a consensus that suggests that MWL can be conceptualized as the interaction between the structure of systems and tasks, on the one hand, and the capabilities, motivation, and state of the human operator, on the other. More specifically, MWL could be defined as the "cost" that an operator incurs as tasks are performed. Given the multidimensional nature of MWL, no single measurement technique can be expected to account for all of the important aspects of it. Current research efforts in measuring MWL use psycho-physiological techniques, such as electroencephalographic, cardiac, ocular, and respiration measures in an attempt to identify and predict MWL levels. Measurement of cardiac activity has been a useful physiological technique employed in the assessment of MWL, both from tonic variations in heart rate and after treatment of the cardiac signal. The MWL depends on the operational conditions and the complexity of the mission (i.e., has to do with the significance of the general task).

The MWL is directly affected by the challenges that a pilot faces, when he/she has to control the vehicle in a complex, heterogeneous, multitask, and often uncertain environment. Such an environment includes numerous different and interrelated concepts of situation awareness: spatial awareness for instrument displays; system awareness (e.g., for keeping the pilot informed about actions that have been taken by automated systems), and task awareness that has to do with the attention and task management. Particularly, the off-normal MWL for a single pilot might be (but does not necessarily have to be) twice as high as the ordinary MWL. As to the HCF, this criterion has been introduced by the author of this book just several years ago. This criterion considers, as indicated above, but might not be limited to, professional experience and qualifications; capabilities and skills; level of training; performance sustainability; ability to concentrate; mature thinking; ability to operate effectively, in a tireless fashion, under pressure, and, if needed, for a long period of time (tolerance to stress); team-player attitude; swiftness in reaction, when necessary; and so on.

The above factors affecting the MWL and possibly having an effect also on HCF values are being studied extensively in today's aviation and aerospace psychology practice, and the available information can be effectively used to evaluate the most likely (ordinary, specified) values of the MWL and HCF criteria. Such an evaluation is, however, beyond the scope of this analysis. We simply assume that these values are deterministic parameters that are known (predetermined) in advance for a particular mission or a situation and for a particular individual. It is noteworthy that the ability to evaluate the "absolute" level of the MWL, important as it might be for a typical (noncomparative) evaluation, is less critical in this approach, which is aimed at the comparative assessments of the likelihood of a casualty, when, because of an unforeseeable accident, the MWL suddenly and significantly changes.

The author realizes that some avionic psychologists might feel that our modeling is a bit simplistic from the avionic psychiatrist point of view. But he does not try to create a comprehensive model. His intent is rather to illustrate how some PRA methods and approaches could be effectively employed to quantify the role of the human factor in situations of the type in question. It should be pointed out also that the DEPDF suggested in this chapter contains time, and is thereby different from the DEPDF suggested in Chapter 6. In this connection, the author wants to emphasize that this function can be structured and introduced in different ways depending on the task, the mission, or a situation one intends to model. The author believes that the suggested model and its generalizations, after appropriate sensitivity analyses are carried out, can be helpful when developing guidelines for personnel training; when choosing the appropriate flight simulation conditions; and/or when there is a need to decide if the existing level of automation and/or the existing navigation instrumentation and equipment are adequate. If not, additional and/or more advanced instrumentation and equipment should be developed and installed. He believes also that the taken approach, with the appropriate modifications and generalizations, is applicable to many other situations, not necessarily in the avionic or even vehicular domain, when a human encounters an uncertain environment and/or a hazardous situation and/or interacts with never perfect hardware and software. It goes without saying that plenty of additional engineering analyses and human psychology–related efforts will be needed, of course, to make the guidelines based on the suggested concept practical and effective.

7.3 Double-Exponential Probability Distribution Function

The following DEPDF

$$Q(F,G) = 1 = \exp\left[-\frac{t}{T}\frac{G^2}{G_0^2}\exp\left(-\frac{F^2}{F_0^2}\right)\right], \tag{7.1}$$

"Two Men in a Cockpit" 165

which is more or less of the EVD type (see, for instance, Refs. [19–22]), can be used to characterize and to quantify, on the comparative basis, the probability $Q(F)$ of a vehicle operator failure to perform his/her duties, because of the mental overload and/or because of insufficient human capacity. In the formula (7.1), $\frac{t}{T}$ is the (nonrandom) ratio of the elapsed operation time, t, to the total duration, T, of the mission (flight) ($0 \le t \le T$), G is the MWL level, G_0 is the most likely (specified, anticipated) value of the MWL for a pilot in ordinary flight conditions, F is the HCF level, and F_0 is its most likely (specified) value. The probability density function can be obtained, if necessary, from (7.1) by differentiation. The formula (7.1) makes physical sense. Indeed, at the beginning of the flight ($t = 0$) and/or when the MWL is very low ($G \to 0$) and/or when the pilot is highly experienced, highly skilled, highly trained, and highly effective ($F \to \infty$), the probability $Q(F)$ that the pilot fails to carry out his/her duties is zero. When the pilot has to operate for an infinitely long time ($t \to \infty$) and/or when the MWL is very high ($G \to 0$), while the HCFF is finite, then the human failure is inevitable: the probability $Q(F)$ is equal to one.

From (7.1), we obtain, by differentiation:

$$\frac{dQ}{dG} = 2\frac{H}{G}, \tag{7.2}$$

where

$$H = -P \ln P \tag{7.3}$$

is the entropy of the probability $P = 1 - Q$ of nonfailure (reliability) of the human to perform his/her duties. As follows from (7.2), the distribution (7.1) reflects a hypothesis that the change in the probability of failure (or nonfailure) with the change in the MWL level G is proportional to the entropy (uncertainty) H of the distribution of this probability and is inversely proportional to the workload level G. When the workload is certain ($H = 0$) and/or is very high ($G \to 0$), the derivative $\frac{dQ}{dG}$ is zero: the probabilities Q and P do not change with the change in the MWL level. When the load G is highly uncertain (large H value) and/or is very low ($G \to 0$), the derivative $\frac{dQ}{dG}$ is significant. This means that the probability of failure is highly dependent on the level of the workload.

The rationale behind the hypothesis (7.2) is that the change in the probability of failure with the change in the level of loading, such as MWL, is proportional to the entropy, the degree of uncertainty, which is assessed as the ratio of the entropy of the distribution of the probability of failure to the level of loading. This ratio could be viewed as a sort of a random coefficient of variation (cov). Unlike the conventional cov (see Chapter 2), which is a nonrandom characteristic defined as the ratio of the standard deviation to the mean value of the random variable of interest, the cov in the right part of the formula (7.2) is a random variable and is defined as the ratio of the (nonrandom) entropy to the (random) level of loading. Both the entropy and the loading level could be time dependent: the "demand" (MWL) and the "capacity" (HCF) distributions might broaden (get "out spread") in time and/or the gap between their most likely values might get narrower when time progresses.

7.4 "Two Men in a Cockpit": "Accident" Occurs When One of the Pilots Fails to Perform His/Her Duties; "Casualty" Occurs If They Both Fail to Do So

It is assumed in this analysis that the two pilots in a cockpit have overlapping duties. There would be more work for the PIC if his/her mate becomes incapacitated; however, it is unclear, of course, to what extent the workload would actually increase. As is known, one of the main reasons for today's two-pilot operation practice is that one pilot can fly the aircraft alone, if needed, and the flight will still be safe. The aircraft and the operation procedures are designed in such a way that this is possible indeed. In this analysis, we assume, however, just to be specific and to demonstrate the attributes of our approach, that the workload doubles as a result of the mishap. The general PRA formalism will remain the same, if the increase in the workload is much lower (which seems to be more likely) or somewhat higher (which is less likely) than by a factor of 2. Consider a situation that takes place at an arbitrary moment t of time after an aircraft takes off for a flight of the anticipated duration T, and assume, for the sake of simplicity, that both pilots are similar (i.e., equally qualified, equally well trained, etc.)—i.e., have equal HCFs—and that, at the beginning of the flight, the MWL G is evenly distributed between them. Using the formula (7.1), one can write the probability of failure for each of the pilots to perform his/her duties in the normal flight conditions as

$$Q_{1/2}(F) = 1 - \exp\left[-\frac{T-t}{T}\frac{G^2}{4G^2}\exp\left(-\frac{F^2}{F_0^2}\right)\right]. \tag{7.4}$$

Note that if it were a reason to believe that the increase in the MWL for the PIC, because of the mishap, is different than two, the factor "4" in the denominator in Equation (7.4) could be changed, so that the uneven distribution of the ordinary MWL is accounted for. If at the moment t of time an

"Two Men in a Cockpit" 167

accident takes place (one of the pilots becomes incapacitated), then his/her mate (PIC) will have to cope with the entire MWL G. The probability that the PIC fails during the remaining time $T - t$ to cope with the increased MWL can be expressed, assuming that the MWL doubles because of the accident, as

$$Q_1(F) = 1 - \exp\left[-\frac{T-t}{T}\frac{G^2}{G_0^2}\exp\left(-\frac{F^2}{F_0^2}\right)\right]. \tag{7.5}$$

The formulas (7.4) and (7.5) indicate that if the accident takes place at the last moment $t - T$ of the flight (including landing), and the workload G is not infinitely large (say, because the environmental conditions are reasonably favorable and the navigation instrumentation and other hardware and software are adequate and reliable), then both probabilities $Q_{0.5}(F)$ and $Q_{1.0}(F)$ are zero, and no casualty could possibly occur: it is simply "too late in the flight" for a mishap to happen. As has been mentioned above, in the current treatment we do not address the "end-of-flight" (landing) situation directly. It is noteworthy that such a situation could be indirectly considered, in a tentative fashion, within the framework of the introduced formalism by simply adequately increasing the hypothetical MWL level in the en route conditions (i.e., by introducing a certain margin of safety). Another, perhaps, better substantiated approach is to consider the contributions of the en route and descending-and-landing situations by introducing a cumulative DEPDF function that would account (with an emphasis on the landing, rather than en route, conditions) for the contributions of both conditions to the cumulative probability of a possible casualty. If the accident takes place at the initial moment $t = 0$ of time (i.e., at the very beginning of the flight), then the formulas (7.4) and (7.5) yield:

$$Q_{1/2}(F) = 1 - \exp\left[-\frac{G^2}{4G_0^2}\exp\left(-\frac{F^2}{F_0^2}\right)\right], \tag{7.6}$$

$$Q_1(F) = 1 - \exp\left[-\frac{G^2}{G_0^2}\exp\left(-\frac{F^2}{F_0^2}\right)\right]. \tag{7.7}$$

If, in such a situation, the total MWL G is high and the HCF F is finite, the probabilities (7.6) and (7.7) are equal to one: the human failure will inevitably take place and the aircraft casualty will certainly occur. If, however, the total MWL G is low, while the HCF F is significant, then these probabilities become zero: no human failure or an aircraft casualty is likely to occur.

7.5 Probability of a Casualty If One of the Pilots Becomes Incapacitated

No casualty could possibly occur if one of the following cases takes place (see Example 2.2):

1. None of the pilots fails to perform his/her duties during the entire flight.
2. The captain fails to perform his/her duties, but the first officer takes over completely and successfully the operation of the aircraft.
3. The first officer fails to perform his/her duties, but the captain takes over completely and successfully the operation of the aircraft.

The probability of the first event is $(1 - Q_{1/2})^2$. The probabilities of the second and the third events are the same and are equal to $Q_{1/2}(1 - Q_1)$. The probability of an accident-free flight is, therefore,

$$P = \left(1 - Q_{1/2}\right)^2 + 2Q_{1/2}\left(1 - Q_1\right) = 1 + Q_{1/2}\left(Q_{1/2} - 2Q_1\right), \tag{7.8}$$

and the probability of a casualty is

$$Q = Q_{1/2}\left(2Q_1 - Q_{1/2}\right). \tag{7.9}$$

From (7.2) and (7.3), we obtain

$$Q_1(F) = 1 - [1 - Q_{1/2}(F)]^4 \tag{7.10}$$

$$Q_{1/2}(F) = 1 - \sqrt[4]{1 - Q_1(F)}. \tag{7.11}$$

If none of the pilots fails—that is, $Q_{1/2}(F) = Q_1(F) = 0$—no accident could possibly occur, and the formula (7.9) yields $Q = 0$. If one of the pilots is unable to cope even with half of the MWL—that is, when $Q_{1/2}(F) = 1$—then, as follows from the formula (7.10), he/she will not be able to cope with the total workload either—that is, $Q_1(F) = 1$—as well, and the casualty becomes inevitable: $Q = 1$. When the probability $Q_{1/2}(F)$ is small, the probability $Q_1(F)$ is also small, and, as evident from (7.10), can be found as $Q_1(F) = 4Q_{1/2}(F)$, so that the probability of the casualty in this case is

$$Q = 7Q_{1/2}^2 \approx \frac{7}{16}Q_1^2 \approx 0.4375Q_1^2.$$

Thus, the probability of a casualty is very low, if the probability that a pilot fails to perform his/her duties in ordinary conditions is small. This

"Two Men in a Cockpit" 169

intuitively obvious circumstance is quantified by the above reasoning. The probabilities $Q_{1/2}(F)$ and $Q(F)$ are computed as functions of the probability $Q_1(F)$ in Table 7.1.

The following conclusions could be drawn from the computed data:

1. The probability $Q_1(F)$ that the pilot fails to cope with the total work-load is always higher than the probability $Q_{1/1}(F)$ of failure to perform his/her normal duties. Table 7.1 data quantify this intuitively obvious fact.

2. The "good news" though is that the probability Q of a casualty is, in general, substantially lower than the probability $Q_{1.0}(F)$ that one of the pilots becomes unable to cope with the total workload. This is especially true when the probability $Q_1(F)$ of his/her ability to handle the entire workload is low. No wonder that there were no casualties in the recently reported accidents, when one of the two pilots became incapacitated during the flight: June 18, 2009, Continental flight 61 from Brussels to Newark; February 24, 2008, GB Airways flight BA6826 from Manchester to Cypress; or January 23, 2007, Continental flight 1838 from Houston to Puerto Vallarta. It is only when the likelihood of the inability of the PIC to handle the entire workload is next to one, then the likelihood of the casualty becomes also close to one. If, for example, one wants to keep the probability of a casualty below, say, $10^{-5} = 0.001\%$, then the probability $Q_1(F)$ that one of the pilots will be unable to cope, if necessary, with the entire workload should be kept below 0.5%. If the latter probability is as high as 10%, then the probability of a casualty becomes as high as 0.45%. These data indicate that a quantitative assessment of the likelihood of failure sheds additional light on the possible consequences of a mishap.

3. The formulas (7.4) and (7.5) and Table 7.1 data indicate that the probability Q of a casualty is lower than the probability $Q_{1/2}(F)$ of a mishap (i.e., the probability of failure of one of the pilots to cope with half of the workload, if the probability Q_1 of his/her failure to cope with the entire workload is below $Q_1 = 1 - \dfrac{1}{2\sqrt[3]{2}} = 0.6031$. The probability Q of a casualty becomes higher than the probability $Q_{1/2}(F)$ of failure of one of the pilots to cope with half of the workload, if the probability Q_1 of his/her failure to cope with the entire workload is higher than the above (rather high and, hence, unrealistic) number. It is clear that there is always a strong incentive to make the probability of failure of each pilot at the ordinary flight conditions as low as possible. That is why the HCF and particularly professional training and work experience are important. Table 7.1 data quantify this intuitively obvious conclusion.

TABLE 7.1

The Probabilities $Q_{1/2}(F)$ of Human Failure to Successfully Handle Half of the MWL F and the Probability Q of Casualty as Functions of the Probability $Q_1(F)$ of Failure

$Q_1(F)$	$Q_{1/2}(F)$	Q
0	0	0
0.000001	2.5E-7	4.375000E-12
0.00001	2.5E-6	4.375000E-11
E-4	2.5E-5	1.750000E-8
5E-4	1.25E-4	1.093750E-7
E-3	2.5E-4	4.375000E-7
5E-3	1.252E-3	1.095250E-5
5E-2	1.2741E-2	1.11177E-3
E-1	2.5996E-2	4.52341E-3
2.0E-1	5.4258E-2	18.75927E-3
3.0E-1	8.5309E-2	4.390777E-2
4.0E-1	11.9888E-2	8.153727E-2
5.0E-1	0.159103	0.13378923
6.0E-1	0.204729	0.20376084
7.0E-1	0.259917	0.29632695
8.0E-1	0.331260	0.42028281
8.5E-1	0.377667	0.49939776
9.0E-1	0.437659	0.59624080
9.5E-1	0.527129	0.72368012
1.0000	1.000000	1.00000000

7.6 Required Extraordinary (Off-Normal) versus Ordinary (Normal) HCF Level

Let us assess, based on the distribution (7.1), how the HCF F should change, so that the probabilities Q_1 and Q are still maintained low despite the occurrence of a mishap. From (7.5), we find

$$\frac{F}{F_0} = \sqrt{-\ln\left[-\frac{T}{T-t}\frac{G_0^2}{G^2}\ln\left(1-Q_1\right)\right]}. \tag{7.12}$$

The worst-case scenario takes place when the accident occurs at the middle of the flight, when there is no appreciable incentive for turning back, and the PIC has to operate the aircraft by himself/herself for the longest time. So, let us assume that the accident took place at the moment t of time, for which $\frac{t}{T} = 0.5$, and that the MWL G, after the mishap has happened, is twice as high as the ordinary (specified) workload G_0. Then the formula (7.12) yields

$$\frac{F}{F_0} = \sqrt{-\ln\left[-\frac{1}{2}\ln\left(1-Q_1\right)\right]}. \tag{7.13}$$

If, for instance, one requires that the probability Q of the casualty does not exceed, say, $Q = 10^{-5}$, then, as follows from Table 7.1 data, the probability Q_1 should be kept below $Q_1 = 0.005 = 0.5\%$, and the formula (7.13) yields $\frac{F}{F_0} = 2.4472$. Hence, the requirements for the HCF, whatever the definition and the structure of this factor are agreed upon and what the corresponding *figures-of-merit* might be, are such that the extraordinary (off-normal) HCF should be by a factor of 2.45 larger than the ordinary (normal) value of this factor. In other words, the pilot should be trained (e.g., on a flight simulator) in such a way that he/she would exhibit/manifest the necessary skills and capabilities in off-normal situations, if/when they occur, by a factor of 2.45 larger than in normal conditions. If one requires that the probability of failure does not exceed a number as low as $Q = 10^{-7}$, then the required $\frac{F}{F_0}$ ratio is higher: $\frac{F}{F_0} = 2.88$. As evident from these data, even such a considerable change in the level of the hazardous condition did not lead to a significant change in the requirements for the HCF level: it is a threefold increase in what seems to be acceptable in normal flight conditions that is sufficient for a satisfactory human performance in a hazardous situation. The practical conclusion from this result is that the existing methods of training pilots, including those on flight simulators, are most likely adequate: no dramatic increase in the mental overload during training is necessary.

Let us consider now a hypothetical situation, when the accident took place at the initial moment of time, but the PIC and the air traffic controller decided, nonetheless, not to interrupt the schedule and to continue the flight. How risky might such a decision be? In this case the formula (7.12) yields the following:

$$\frac{F}{F_0} = \sqrt{-\ln\left[-\frac{1}{4}\ln\left(1-Q_1\right)\right]}. \tag{7.14}$$

For probabilities $Q = 10^{-5}$ and $Q = 10^{-7}$ of the likely casualty, we obtain $\frac{F}{F_0} = 2.5850$ and $\frac{F}{F_0} = 2.9978$, respectively. Hence, the time of an accident has a relatively small effect on the required increase in the level of the off-normal HCF, so that the threefold increase compared to normal conditions seems to be still adequate. Thus, the hypothetical decision to continue the flight, instead of returning to the airport of departure, might not be unjustifiable. We would like to point out that the above factors of the increased MWL during training, although they are a logical output of the employed probabilistic formalism, might have different interpretation in today's actual practice, when skilled and qualified pilots are trained to the best of the ability of the existing simulators and personnel involved. The author of this book simply states that, based on the obtained probabilistic information, the MWL during training, whatever its level is, does not have to be infinitely and indefinitely high, but could be "just" ("only") threefold more extensive than the regular expected MWL. It would be extremely useful, of course, if it would be possible to establish a quantitative measure of how high the typical MWL level is during training (i.e., when the pilots are prepared to operate effectively both in ordinary and extraordinary conditions) in comparison with ordinary flight conditions. This effort is, however, beyond the scope of our chapter. It would be highly desirable to carry out such an analysis on a flight simulator. In a way, such testing/training is similar to highly accelerated life testing in electronics and optical engineering.

7.7 Probabilistic Assessment of the Effect of the Time of Mishap on the Likelihood of Accident

The objective of the analysis carried out in this section is to assess, based on the developed formalism, the effect of the time of operation on the likelihood of a casualty. Let the probability Q of a possible casualty be evenly distributed over the time T of the flight, and no casualty has actually occurred during the initial time t of the flight. What is the probability Q_* that the casualty will occur during the remaining time $T - t$ (see also Example 2.3)?

Two events have to take place for the accident to occur during the time $T - t$: (1) the accident should not take place during the time t and (2) has to occur during the time $T - t$. The probability that the casualty occurs during the time t is $Q\dfrac{t}{T}$, and the probability that the casualty occurs during the remaining time $T - t$, provided that it did not occur during the time t, is $\left(1 - Q\dfrac{t}{T}\right)Q_*$. The probability Q that the casualty occurs during the total time T is

$$Q = Q\frac{t}{T} + \left(1 - Q\frac{t}{T}\right)Q_*. \tag{7.15}$$

Hence, the probability Q_* that the casualty will occur during the remaining time $T - t$ is related to the probability Q of the occurrence of the accident during the entire flight (mission) as follows:

"Two Men in a Cockpit"

$$Q_* = Q \frac{1 - \dfrac{t}{T}}{1 - Q\dfrac{t}{T}}. \tag{7.16}$$

Hence,

$$Q = \frac{Q_*}{1 - \dfrac{(1 - Q_*)t}{T}}. \tag{7.17}$$

The computed Q_* values versus the time ratio $\dfrac{t}{T}$ are shown in Table 7.2.

The following major conclusions could be drawn from the computed data:

1. The probability Q_* that the casualty occurs during the remaining time $T - t$ of the flight, if it did not occur during the time t, is always smaller than the specified probability Q of the casualty occurrence during the total flight time T, and decreases with an increase in the total time T of the flight.
2. At the last moment $t = T$. of time the probability Q_* is, of course, zero, no matter how high the probability Q is, unless the latter probability is equal to one.

TABLE 7.2

The Computed Probabilities Q_* That the Casualty Will Occur during the Remaining Time as a Function of This Time and the Probability That the Casualty Occurs at This Flight

$\dfrac{t}{T}Q$	0	0.2	0.4	0.6	0.8	1.0
0	0	0	0	0	0	0
E-7	E-7	8E-8	6E-8	4E-8	2E-8	0
E-5	E-5	8E-6	6E-6	4E-6	2E-6	0
E-3	E-3	8E-4	6E-4	4E-4	2E-4	0
0.01	0.01	0.008	0.006	0.004	0.002	0
0.04	0.04	0.0323	0.0244	0.0164	0.00826	0
0.08	0.08	0.0650	0.0522	0.0336	0.0171	0
0.10	0.10	0.0816	0.0625	0.0420	0.0217	0
0.20	0.20	0.1667	0.1304	0.0909	0.0476	0
0.40	0.40	0.3478	0.2857	0.2105	0.1176	0
0.60	0.60	0.5455	0.4737	0.3750	0.2308	0
0.80	0.80	0.7619	0.7059	0.6154	0.4444	0
1.00	1.00	1.00	1.00	1.00	1.00	–

3. The probability Q_* that the accident will occur during the remaining time $T - t$ increases with an increase in the specified probability Q. The two probabilities coincide at the initial moment of time $t = 0$.

4. If one wants to keep the probability Q_* at a sufficiently low level, then he/she should keep the specified probability Q also at a low level.

5. The above intuitively more or less obvious conclusions are quantified by Table 7.2 data.

7.8 Conclusions

The following conclusions can be drawn from the performed analysis:

- A DEPDF is introduced to characterize and to quantify the likelihood of a human failure to perform his/her duties when operating a vehicle (a car, an aircraft, a boat, etc.). This function is applied to a particular safety-in-air situation, when one of the two pilots becomes, for whatever reason, incapacitated at a certain moment of time in the flight, and, because of that, the other pilot has to cope with a considerably higher MWL as compared to regular flight conditions. The probability of this event is assessed as a function of the HCF.

- It is shown how methods of the classical probability theory could be effectively employed to quantify the role of the human factor in the situation in question. The suggested model, after an appropriate sensitivity analysis is carried out, might be used particularly when developing guidelines for personnel training and/or when there is a need to decide if the existing navigation instrumentation is adequate in extraordinary safety-in-air situations, or if additional and/or more advanced equipment should be developed and installed.

- The numerical data based on the suggested model make physical sense and are in satisfactory, at least, qualitative, agreement with the existing practice. In conclusion, we would like to point out that it is important to relate the model expressed by the basic equation (7.1) to the existing practice, and to review the existing practice from the standpoint of the, generally, consistent model expressed by this equation.

- Further research, refinement, and validation would be needed, of course, before the model could see practical application.

References

1. D.C. Foyle and B.L. Hooey, *Human Performance Modeling in Aviation*, CRC Press, Boca Raton, FL, 2008.
2. S. Svensson, M. Angelborg-Thanderz, and L. Sjoeberg, "Mission Challenge, Mental Workload and Performance in Miliary Aviation," *Aviation, Space, and Environmental Medicine*, vol. 64, No. 11, 1993, pp. 985–991.
3. T.C. Hankins and G.F. Wilson, "A Comparison of Heart Rate, Eye Activity, EEG, and Subjective Measures of Pilot Mental Workload during Flight," *Aviation, Space, and Environmental Medicine*, vol. 69, 1998, pp. 360–367.
4. K.A. Greene, K.W. Bauer, M. Kabrisky, S.K. Rogers, and G.F. Wilson, "Estimating Pilot Workload Using Elman Recurrent Neural Networks: A Preliminary Investigation." In C.H. Dagli, et al. (eds.), *Intelligent Engineering Systems through Artificial Neural Networks*, vol. 7, ASME Press, New York, November 1997.
5. J.A. East, K.W. Bauer, and J.W. Lanning, "Feature Selection for Predicting Pilot Mental Workload: A Feasibility Study," *International Journal of Smart Engineering System Design*, vol. 4, 2002, pp. 183–193.
6. J.B. Noel, "Pilot Mental Workload Calibration," MS Thesis, School of Engineering, Air Force Institute of Technology, Wright-Patterson AFB, OH, March 2001.
7. A.W. Gaillard, "Comparing the Concepts of Mental Load and Stress," *Ergonomics*, vol. 36, 1993, pp. 991–1005.
8. A.F. Kramer, "Physiological Metrics of Mental Workload: A Review of Recent Progress." In P. Ullsperger (ed.), *Mental Workload*, Bundesanstalt für Arbeitmedizin, Berlin, 1993, pp. 2–34.
9. L.R. Fournier, G.F. Wilson, and C.R. Swain, "Electrophysiological, Behavioral and Subjective Indexes of Workload When Performing Multiple Tasks: Manipulation of Task Difficulty and Training," *International Journal of Psychophysiology*, vol. 31, 1999, pp. 129–145.
10. P. Ullsperger, A. Metz, and H. Gille, "The P300 Component of the Event-Related Brain Potential and Mental Effort," *Ergonomics*, vol. 31, 1988, pp. 1127–1137.
11. G.F. Wilson, P. Fullerkamp, and I. Davis, "Evoked Potential, Cardiac, Blink and Respiration Measures of Pilot's Workload in Air-to-Ground Missions," *Aviation, Space and Environmental Medicine*, vol. 65, 1994, pp. 100–105.
12. J.B. Brookings, G.F. Wilson, and C.R. Swain, "Psychophysiological Responses to Changes in Workload during Simulated Air Traffic Control," *Biological Psychology*, vol. 42, 1996, pp. 361–677.
13. T.C. Hankins and G.F. Wilson, "A Comparison of Heart Rate, Eye Activity EEG and Subjective Measures of Pilot Mental Workload during Flight," *Aviation, Space and Environmental Medicine*, vol. 69, 1998, pp. 360–367.
14. A.H. Roscoe, "Heart Rate as a Psycho-Physiological Measure for In-flight Workload Assessment," *Ergonomics*, vol. 36, 1993, pp. 1055–1062.
15. R.W. Backs, "Going beyond Heart Rate: Autonomic Space and Cardiovascular Assessment of Mental Workload," *International Journal of Aviation Psychology*, vol. 5, 1995, pp. 25–48.

16. A.J. Tattersall and G.R. Hockey, "Level of Operator Control and Changes in Heart Rate Variability during Simulated Flight Maintenance," *Human Factors*, vol. 37, 1995, pp. 682–698.
17. P.G. Jorna, "Heart Rate and Workload Variations in Actual and Simulated Flight," *Ergonomics*, vol. 36, 1993, pp. 1043–1054.
18. J.A. Veltman and A.W. Gaillard, "Physiological Indices of Workload in a Simulated Flight Task," *Biological Psychology*, vol. 42, 1996, pp. 323–342.
19. E. Suhir, *Applied Probability for Engineers and Scientists*, McGraw-Hill, New York, 1997.
20. E. Suhir, "Adequate Underkeel Clearance (UKC) for a Ship Passing a Shallow Waterway: Application of the Extreme Value Distribution (EVD)," OMAE, Rio-de-Janeiro, Brazil, 2001.
21. E. Suhir, "Helicopter Landing Ship (HLS): Undercarriage Strength and the Role of the Human Factor," Honolulu, Hawaii, OMAE 2009, see also *ASME Transactions, Journal of Offshore Mechanics and Arctic Engineering* (OMAE), vol. 132, February 2010.
22. E. Suhir, "Likelihood of a Vehicular Mission Success and Safety, and the Role of the Human Factor," Paper #1168, IEEE Aerospace Conference, Big Sky, MT, March 2011.
23. E. Suhir, "How to Make a Device into a Product: Accelerated Life Testing, Its Role, Attributes, Challenges, Pitfalls and Interaction with Qualification Testing." In E. Suhir, C.P. Wong, and Y.C. Lee (eds.), *Micro- and Opto-Electronic Materials and Structures: Physics, Mechanics, Design, Packaging, Reliability*, vol. 2, Chapter 8, Springer, 2007, pp. 203–232.

8

Probabilistic Modeling of the Concept of Anticipation in Aviation

8.1 Summary

Considerable improvements in safety-in-the-air situations can be expected through better ergonomics, better work environments, and other efforts that directly affect human behavior and performance. There is also a significant potential for reducing aeronautic casualties through better understanding the roles that various uncertainties of different natures play in the operator's world of work. The probabilistic concept proceeds from the understanding that uncertainties are an important feature of human behavior and performance, that "nobody and nothing is perfect," and that the difference between a success and a failure in a particular effort, situation, or mission is, in effect, "merely" the difference in the levels of the never-zero probability of failure. In this analysis, two problems that have to do with uncertainties in an anticipation effort as an important cognitive resource for improved aeronautics safety are addressed on the probabilistic basis using analytical (mathematical) modeling: (1) evaluation of the probability that the random "subjective" ("internal," pilot performance related) anticipation time is below the random "objective" ("external," "available," situation related) time in the anticipation process; and (2) assessment of the likelihood of success of the random short-term (extraordinary, off-normal) anticipation from the predetermined deterministic (ordinary, normal) long-term anticipation. The employed formalisms, with the appropriate modifications and generalizations, are applicable not only in the avionic and even not only in the vehicular domain, but in many other human performance-related situations, when a human encounters an uncertain and often harsh environment and has to make a decision under a more or less significant time pressure.

177

8.2 Introduction

Substantial improvement in safety-in-the-air can be expected through better ergonomics, better work environment, and other aspects that directly affect human behavior and performance (see, e.g., Refs. [1–3]). There is also a potential for reducing avionic casualties through better understanding the role that various uncertainties of different natures play in the pilot's world of work [4–12]. Such uncertainties might be associated with the human factor directly (human fatigue, delayed reaction, inadequate anticipation and situation awareness, erroneous judgment, and/or wrong decision-making, etc.) and/or indirectly, because of the imperfect forecast or as a consequence of pilot's interaction with various never-perfect instrumentation and equipment, such as, for example, unreliable hardware and software; variable and often harsh environmental conditions; insufficient accuracy and user friendliness of information; predictability and timeliness of the response of the aircraft to the pilot's actions, and so on. By employing quantified (measurable) ways of assessing the role of these uncertainties ("nobody and nothing is perfect"), one could improve dramatically the human performance, when a human becomes part (often the most crucial part) of a complex "man-instrumentation-aircraft-environment" system. Such a probabilistic risk analysis (PRA) based and "human-in-the-loop" (HITL) focused approach enables one to consider, predict, quantify, and, if necessary, minimize and even specify the probability of the occurrence of a mishap. Not all the uncertainties have to be considered, of course, in every safety-in-the-air and HITL-related situation, and not all the uncertainties have to be accounted for on the probabilistic basis, but it is always imperative that a careful insight into the role of possible critical uncertainties and ability to quantify them are pursued whenever possible and appropriate. In the analysis that follows, we address a critical safety-in-the-air aspect of the human factor—anticipation effort, a cognitive resource for improved aeronautics safety [13–16]. Our objective is to show how human factor could be successfully predicted and quantified on the probabilistic basis. We do not intend to come up with an accurate, ready-to-go, "off-the-shelf"-type and complete methodology, in which all the i's are dotted and all the t's are crossed, but rather intend to demonstrate that an attempt to quantify the role of a human factor in safety-in-the-air situations, when anticipation is crucial, can be fruitful, successful, and practical.

8.3 Anticipation, Its Role and Attributes

Amalberti [17] attributes the mental workload (MWL), when addressing cognitive anticipation stress, to the difference between the actual (real-life) anticipation and planning: while planning aims at designing solutions in order to lead one's activity, anticipation aims at testing this solution using assessments of situation's evolution. Michon [18] supports the idea of a dual nature of time in the anticipation process: "time is the conscious experiential product of the processes that allow the (human) organism to adaptively organize itself so that its behaviour remains tuned to the sequential (i.e., order) relations in its environment." The first aspect of time is implicit. It is an unconscious mental representation of how to tune one's behavior to the environment dynamics. The other aspect is the explicit, objective time, which allows an assessment of its duration. Time can thus be considered as a form of high-level cognition. Friedman [19] shows that time experience lays on cognitive skills, which are, in effect, the ability to assess its evolution and positioning of events within it. Two anticipation-related problems in aeronautics have been considered in this paper with an objective to quantify anticipation on the probabilistic basis.

One problem has to do with the duration of the actual, real-time, anticipation effort as compared to the "available" time (i.e., time until the anticipated event of important commences). While anticipation is defined differently in different fields of human psychology [13], we proceed, following Cellier [20], from the definition that anticipation is "an activity consisting of evaluating the future state of a dynamic process, determining the time and timing of actions to undertake on the basis of a representation of the process in the future and, finally, mentally evaluating the possibilities of these actions." Thus, in accordance with this definition, one has to assess the (random) durations of the following three time periods affecting the success of the anticipation effort: (1) time required to evaluate the future state of the dynamic process of interest (what will most likely happen, if I do not interfere?); (2) time required to determine the time when the pilot's actions should start and what kind of actions should be taken (when should I start acting, and what exactly should I do in view of what might happen if I do not act?); and (3) time required to determine, by mental evaluation, whether the required actions are possible (are the actions that I intend to undertake possible, and, if they are, will I achieve my objective?). If the likelihood (probability) that the total anticipation time will be appreciably below the moment of time when the anticipated situation in the dynamic process of importance is expected to commence is high, then there is a reason to believe that the anticipation effort will be successful.

Another problem addressed is the probabilistic assessment of the success of a short-term anticipation from the known (predetermined and deterministic) long-term anticipation. When solving this problem, we proceed from Denecker's [9] definition and distinction between short-term ("subsymbolic") and long-term ("symbolic") anticipations. According to Denecker, short-term anticipation (STA) relies on reflex loops and is "a low-level action" control activity, while long-term anticipation (LTA) relies on the solutions based on the accumulated and analyzed knowledge of the situation of interest and the required adequate modus operandi. Control level models such as Amalberti and Hoc's and Hollnagel's [21,22], point out the relationship between control level and temporal depth: the higher the control level the deeper in time one can project oneself. An implementation of knowledge on a short time period or the use of skills in the long term results in an increased MWL. Thus, previous works [23] show that given a low control level, an increase of anticipation time depth increases the MWL. Hancock and Caird [24] postulate that the MWL level is inversely proportional to the perceived effective time for action ("available" time) and is proportional to the perceived distance from goal. In other words, MWL is lower if one assesses a high amount of available time to act with a low perceived distance to the pursued goal. A distant goal or a low amount of available time for action will lead to a higher MWL.

In a dynamic environment, dealing with uncertainty is mandatory. A previous work on anticipation model [13] put the emphasis on this idea: the further from the goal, the higher is the uncertainty. In order to achieve a defined goal, high-level control layers define both the performance-to-risk ratio, preactivate a set of solutions in order to do so, and define the criticality of subobjectives. Each solution has its duration and an intrinsic sensitivity to environment uncertainty, which will make it more or less likely to reach the objective in the predefined terms. Significant ("deep") time projection makes it impossible to deal with uncertainty: because of combinatory explosion, vague assessments of how the situation evolves will be made, which will require fulfilling the situation representation through time. This process is rather economical: it allows spreading MWL until acting. But, implementing knowledge is costly, and this mental effort will be higher if concentrated on a short time period: dealing with high-level uncertainty on a very STA is highly costly. Clearly, implementing knowledge enables to make a long-term projection on a next-to-deterministic basis (i.e., with a very low risk of failure), while a short-term projection requires skills and the ability to act swiftly and adequately in often unpredictable and extraordinary situations [13–21]. The analyses carried out in this paper are conducted within the framework of the general Human Engineering for Aerospace Laboratory (HEAL) effort, and particularly as part of the Anticipation Support for Aeronautical Planning (ASAP) project [13–21]. The goal of the project is to help pilots to better use the cockpit recourses to facilitate and improve their anticipation skills and practice. The taken approach, with the appropriate modifications and generalizations, is applicable to many other situations, not only in the vehicular domain, when a human encounters an uncertain and harsh environment and/or any other hazardous situation and/or interacts with never-perfect instrumentation and equipment.

8.4 Probabilistic Assessment of the Anticipation Time

8.4.1 Probability That the Anticipation Time Exceeds a Certain Level

We assume that the times required to evaluate the state of the approaching of the dynamic process of interest, the time required to determine the moment of action, and the time to decide what kind of actions should be undertaken could be combined into the phase 1 (evaluation phase) of the anticipation time, while the time required to determine, by mental evaluation, whether the required actions are indeed possible are viewed as the phase 2 of the anticipation time (possibility assessment phase). Such a breakdown seems to be justified, since in reality the pilot, in his/her cognitive evaluation of the situation, anticipates most likely concurrently the activities associated with the assessment of the significance and the attributes of the future state of the dynamic process and the moment of time that, after the future state of the dynamic process of interest is established, the pilot says "Now!". Time required to determine, by mental evaluation, whether the actions decided upon are indeed possible and will meet the objective comprises the phase 2 (possibility assessment phase) of the anticipation process. If, for one reason or another, one decides on a different breakdown of the anticipation phases, the formalism developed in the following text is still applicable. If the (random) sum $T = t + \theta$ of the (random) time, t, needed for the completion of the evaluation phase 1 of the anticipation process and the (random) time, θ, of the possibility assessment phase 2 is lower, with a high enough probability, than the "external," available (random) time, L, from the beginning of the anticipation process to the beginning of the dynamic process of interest, then the anticipation process could be considered successful.

Let us assume the simplest, but physically meaningful, probability distributions for the random times of interest. Rayleigh's laws

$$f_t(t) = \frac{t}{t_0^2} \exp\left(-\frac{t^2}{2t_0^2}\right), \quad f_\theta(t) = \frac{\theta}{\theta_0^2} \exp\left(-\frac{\theta}{2\theta_0^2}\right), \tag{8.1}$$

can be selected as a suitable probability density distribution to characterize the random times t and θ. The rationale behind such an assumption is as follows. The times t and θ cannot be negative, the likelihood of zero random times t and θ is zero, and so is the likelihood of their very large values. In addition, and the most likely times and of the random times t and θ should be small enough, and, in any event, should be much closer to zero than to very large values. In the distributions (8.1), and are the maximum values of these distributions and, hence, the most likely values of the random times t

and θ. The mean times $\bar{t}$ and $\bar{\theta}$ of the variables t and θ are related to the most likely times t_0 and θ_0 as

$$\bar{t} = \sqrt{2\pi t_0}, \quad \bar{\theta} = \sqrt{2\pi \theta_0}. \tag{8.2}$$

The (random) time, L, from the beginning of the anticipation process to the beginning of the dynamic process (event) of importance, has a different physical nature than the anticipation times t and θ. While the times t and θ are "subjective" times that have to do with the swiftness and quality of human anticipation, the random time L is an "objective" ("external," "available") time that is independent of the human anticipation. It is natural to assume that the normal law

$$f_1(l) = \frac{1}{\sqrt{2\pi}\sigma} \exp\left(1 - \frac{(l-l_0)^2}{2\sigma_0^2}\right), \quad \frac{l_0}{\sigma_0} \geq 4.0, \tag{8.3}$$

can be used as a suitable approximation for the time, L. In the formulas (8.3), l_0 is the most likely (and also the mean and the median) value of the "external" time L, and σ_0 is its standard deviation. The ratio $\dfrac{l_0}{\sigma_0}$ ("safety factor") of the mean value l_0 of the available time L to its standard deviation σ_0 should be large enough (say, larger than 4), in order that the normal law could be used as an acceptable approximation for a random variable that cannot be negative, and it is indeed the case in question, when this variable is time.

The probability P_* that the sum $T = t + \theta$ of the random variables t and θ (total anticipation time) exceeds a certain time duration (level) $\hat{T}$ can be found as a convolution of the distributions of the random variables t and θ as follows:

$$P_* = 1 - \int_0^{\hat{T}} \frac{t}{t_0^2} \exp\left(-\frac{t^2}{2t_0^2}\right)\left[1 - \exp\left(-\frac{(T-t)^2}{2\theta_0^2}\right)\right]dt = \exp\left(-\frac{\hat{T}^2}{2t_0^2}\right)$$

$$+\exp\left[-\frac{\hat{T}^2}{2\left(t_0^2 + \theta_0^2\right)}\right]\left\{\frac{\theta_0^2}{t_0^2 + \theta_0^2}\left[\exp\left[-\frac{t_0^2\hat{T}^2}{2\theta_0^2\left(t_0^2 + \theta_0^2\right)}\right]\right] - \exp\left[-\frac{\theta_0^2\hat{T}^2}{2t_0^2\left(t_0^2 + \theta_0^2\right)}\right]\right\}+$$

$$+\sqrt{\frac{\pi}{2}}\frac{\hat{T}t_0\theta_0}{\left(t_0^2 + \theta_0^2\right)^{3/2}}\exp\left[-\frac{\hat{T}^2}{2\left(t_0^2 + \theta_0^2\right)}\right]\left\{\Phi\left[\frac{t_0\hat{T}}{\theta_0\sqrt{2\left(t_0^2 + \theta_0^2\right)}}\right] + \Phi\left[\frac{\theta_0\hat{T}}{t_0\sqrt{2\left(t_0^2 + \theta_0^2\right)}}\right]\right\},$$

$$\tag{8.4}$$

Probabilistic Modeling of the Concept of Anticipation in Aviation 183

where

$$\Phi(x) = \frac{2}{\sqrt{\pi}} \int_0^x e^{-z^2} dz \qquad (8.5)$$

is the error function (see, for instance, Ref. [4]).When the time $\hat{T}$ is zero, this time will be always exceeded ($P_* = 1$). When the time $\hat{T}$ is infinitely long $(\hat{T} \to \infty)$, the probability that this time is exceeded is always zero ($P_* = 0$).

When the most likely duration θ_0 of the phase 2 of anticipation is very small compared to the most likely duration, t_0, of the phase 1, the expression (8.4) yields

$$P_* = \exp\left(-\frac{\hat{T}^2}{2t_0^2}\right), \qquad (8.6)$$

that is, the probability that the total anticipation time exceeds a certain time duration, $\hat{T}$, depends only on the most likely time, t_0, of the first phase. From (8.6), we have

$$\frac{t_0}{\hat{T}} = \frac{1}{\sqrt{-2\ln P_*}}. \qquad (8.7)$$

If the acceptable probability P_* of exceeding the time $\hat{T}$ (e.g., the duration of the available time, if this duration is treated as a nonrandom variable of the level $\hat{T}$), is, say, $P = 10^{-4}$, then the anticipation time should not exceed $0.2330 = 23.3\%$ of the time $\hat{T}$ (expected duration of the available time), otherwise the requirement $P \leq 10^{-4}$ will be compromised. Similarly, when the most likely duration, t_0, of the phase 1 of anticipation effort is very small compared to the most likely time, θ_0, of the second phase, the formula (8.4) yields

$$P_* = \exp\left(-\frac{\hat{T}^2}{2\theta_0^2}\right), \qquad (8.8)$$

(i.e., the probability of exceeding the time level $\hat{T}$ depends only on the most likely time, θ_0, of the second phase of anticipation). The time level $\hat{T}$ can be considered as the conscious experience of this tuning process.

As follows from the formulas (8.1), the probability that the duration of the phase 1 or the phase 2 of anticipation time exceeds the corresponding most likely times is expressed by the formulas of the types (8.6) and (8.8), and is as high as $P_* = \frac{1}{\sqrt{e}} = 0.6065$. In this connection it should be pointed out that the one-parametric Rayleigh distribution is characterized by a rather large standard deviation and therefore might be overconservative. A more flexible two-parametric law, such as, for example, the Weibull law, might be more suitable and more practical as an appropriate probability distribution of the random times, t and θ. Its use, however, will make our analysis unnecessarily more complicated.

As has been indicated above, our goal is not to provide a comprehensive and an accurate methodology for the assessment of the success of the anticipation process, but to demonstrate, based on simple and physically meaningful examples, that the attempt to use methods of the probabilistic risk analysis to quantify the role of the human factor in the safety-in-the-air problem in question might be insightful and fruitful. When developing practical guidelines and recommendations, a particular law of the probability distribution should be established based on the actual statistical data, and employment of various goodness-of-fit criteria might be necessary to conduct detailed statistical analyses. Such an effort is, however, beyond the scope of the analysis.

When the most likely times t_0 and θ_0 required to complete the two phases of the anticipation effort are equal, the formula (8.4) yields

$$P_* = P_*\left(\frac{t_0}{\hat{T}}, \frac{\theta_0}{\hat{T}}\right) = \exp\left(-\frac{\hat{T}^2}{2t_0^2}\right)\left[1 + \sqrt{\pi}\,\frac{\hat{T}}{2t_0}\exp\left(\left(\frac{\hat{T}}{2t_0}\right)^2\right)erf\left(\frac{\hat{T}}{2t_0}\right)\right]. \quad (8.9)$$

For large enough $\dfrac{\hat{T}}{t_0}$ ratios $\left(\dfrac{\hat{T}}{t_0} \geq 3\right)$, the second term in the brackets becomes large compared to unity, so that only this term should be considered. The calculated probabilities of exceeding a certain time level $\hat{T}$ based on the formula (8.9), are shown in Table 8.1. In the third row of this table we indicate, for the sake of comparison, the probabilities, P^0, of exceeding the given time, $\hat{T}$, when only the time t_0 or only the time θ_0 is different from zero (i.e., for the

TABLE 8.1

The Probability P_* That the Anticipation Time Exceeds a Certain Time Level $\hat{T}$ versus the Ratio $\hat{T}/t_0$ of This Time Level to the Most Likely Time t_0 of Anticipation for the Case When the Most Likely Time t_0 of the First Phase and the Most Likely Time θ_0 of the Second Phase Are the Same

$\hat{T}/t_0$	6	5	4	3	2
P_*	6.562E-4	8.553E-3	6.495E-2	1.914E-1	6.837E-1
P^0	1.523E-8	0.373E-5	0.335E-3	1.111E-2	1.353E-1
$\dfrac{P_*}{P^0}$	4.309E4	2.293E3	1.939E2	1.723E1	5.053

Note: For the sake of comparison, the probability P^0 of exceeding the time $\hat{T}$, when either the time t_0 or the time θ_0 are zero, is also indicated.

Probabilistic Modeling of the Concept of Anticipation in Aviation **185**

special case that is mostly remote from the case $t_0 = \theta_0$, when the most likely times of the two phases are equal). Clearly, the probabilities computed for other possible combinations of the times t_0 and θ_0 could be found between the calculated probabilities P_* and P^0. The following practically important conclusions could be drawn from Table 8.1 data:

1. The probability that the total time of anticipation exceeds the given time level $\hat{T}$ rapidly increases with an increase in the time of anticipation.

2. The probability of exceeding the time level $\hat{T}$ is considerably higher, when the most likely times of the two phases of anticipation time are finite, and particularly are equal to each other, in comparison with the situation when one of these times is significantly shorter than the other (i.e., zero or next to zero). This is especially true for short anticipation times: the ratio $\dfrac{P_*}{P^0}$ of the probability P_* of exceeding the time level $\hat{T}$ in the case of $t_0 = \theta_0$ to the probability P^0 of exceeding this level in the case $t_0 = 0$ or in the case $\theta_0 = 0$ decreases rapidly with an increase in the duration of anticipation time. Thus, a significant incentive exists for reducing the total anticipation time. This intuitively obvious fact is quantitatively assessed in our analysis.

The data of the type shown in Table 8.1 can be used, particularly, to train the personnel for a quick reaction, as far as the anticipation process is concerned. If, for example, the expected duration of the available time is 30 s, and the required (specified) probability of exceeding this time is $P = 10^{-3}$ (0.1%), then, as evident from the table data, the times for each of the two phases of the anticipation process should not exceed 5.04 s. It is advisable, of course, that these predictions are verified by simulation and by actual best practices. Particularly, one should obtain statistical information, from the accumulated experience, about the available time durations for different practical situations. Other useful information that could be drawn from the data of the type shown in Table 8.1 is whether it is possible at all to train a human to react (make a quick and reasonable anticipation) in just several seconds. If not, then one should decide on a broader involvement of more sophisticated, more powerful, and more expensive equipment to do the job. If pursuing such an effort is decided upon, then probabilistic sensitivity analyses will be needed to determine the most promising ways to go.

8.4.2 Probability That the Duration of the Anticipation Process Exceeds the Available Time

The available time L is a random normally distributed variable, and the probability that this time is found below a certain level $\hat{L}$ can be determined as

$$P_l = P_l\left(\frac{\sigma_0}{\hat{L}},\frac{l_0}{\hat{L}}\right) = \int_{-\infty}^{\hat{L}} f_l(l)\,dl = \frac{1}{2}\left[1+\Phi\left(\frac{\hat{L}-l_0}{\sqrt{2}\sigma_0}\right)\right]$$

$$= \frac{1}{2}\left[1+\Phi\left(\frac{1-\dfrac{l_0}{\hat{L}}}{\sqrt{2}\,\dfrac{\sigma_0}{\hat{L}}}\right)\right]. \tag{8.10}$$

The probability that the available time in the anticipation situation is exceeded can be determined by equating the times $\hat{T} = \hat{L} = T$ and computing the product

$$P_A = P_*\left(\frac{t_0}{T},\frac{\theta_0}{T}\right)P_l\left(\frac{\sigma_0}{T},\frac{l_0}{T}\right) \tag{8.11}$$

of the probability $P_*\left(\dfrac{t_0}{T},\dfrac{\theta_0}{T}\right)$ that the total time of anticipation exceeds a certain level, T, and the probability $P_l\left(\dfrac{\sigma_0}{T},\dfrac{l_0}{T}\right)$ that the duration of the available time is shorter than the time T. The formula (8.11) considers the effect of the "objective" situation (through the values of the most likely duration, l_0, of the random available time, L, and its standard deviation σ_0), the role of the human factors t_0 and θ_0 (the most likely times of the anticipation process phases; these times characterize the pilot qualifications) on the probability of the success of the anticipation process. After a low enough acceptable value P_A^* of the probability P_A is established (agreed upon), Equation (8.11) can be used to establish the allowable maximum most likely time θ_0 of the second phase of the anticipation process. The actual time of the second (final) phase of the anticipation process can be assessed by the formula of the type (8.7):

$$\Delta t^* = \theta_0\sqrt{-2\ln P_l}, \tag{8.12}$$

where P_l is the allowable probability that the level Δt^* is exceeded. If, for instance, $\theta_0 = 10$ s and $P_l = 10^{-5}$, then $\Delta t^* = 48.0$ s.

Let the most likely times of the two phases of the anticipation process be the same and equal to $t_0 = \theta_0 = 10$ s, the most likely (mean) available time be $l_0 = 20$ s, and the standard deviation of the available time be $\sigma_0 = 5$ s. Then, using the formulas (8.10) and (8.11), and Table 8.1 data, we obtain the data shown in Table 8.2.

As evident from Table 8.2 data, the probability P_A that the total anticipation time exceeds the duration of the available time (failure of the anticipation process) increases rapidly with the decrease in the ratio of the duration of the available time to the most likely time of either of the two phases of the

Probabilistic Modeling of the Concept of Anticipation in Aviation

TABLE 8.2

The Probability P_A of the Success of the Anticipation Effort versus the Ratio T/t_0 of the Normally Distributed Duration T of the "External" Time to the Most Likely Time t_0 of the First Phase of the Anticipation Process or the Most Likely Time θ_0 of the Second Phase of This Process, When the Times t_0 and θ_0 Are Equal

T/t_0	6	5	4	3	2
P_*	6.562E-4	8.553E-3	6.495E-2	1.914E-1	6.837E-1
T/l_0	3.0	2.5	2.0	1.5	1.0
P_l	1.0	1.0	0.9999	0.9770	0.5000
P_A	6.562E-4	8.553E-3	6.494E-2	1.870E-1	3.418E-1

anticipation effort, while the probability that the available time is below a certain value, decreases with the decrease in the ratio of this value to the most likely duration of the available time. The first effect prevails, and the product of these two probabilities (defining the likelihood that the anticipation effort fails) increases with the decrease in the duration of the available time almost as fast as the probability of the anticipation time does. It is only for very long anticipation times that the probability P_l of exceeding a certain time limit starts to play an appreciable role. We conclude, therefore, that in the situation in question, the human factor associated with the anticipation times plays a significant role, as far as the success of the anticipation effort is concerned. The developed model enables one to quantitatively assess in the problem in question this role, along with other uncertainty sources. The success of the anticipation effort can be expected if the probability that it takes place during the available time is sufficiently high. The developed simple and easy-to-use formulas enable one to evaluate this probability. The model can be used particularly when developing guidelines for personnel training. Plenty of additional risk analyses and human psychology-related effort will be needed, of course, to make such guidelines practical.

8.5 Probabilistic Assessment of the Success of STA from the Predetermined LTA

8.5.1 Double-Exponential Probability Distribution Function

According to Denecker [19], STA relies on reflex loops and is "a low level action" control activity. LTA relies on the solutions based on the accumulated and analyzed knowledge of the situation of interest and the required adequate modus operandi. Implementing knowledge enables one to make a long-term projection on a next-to-deterministic basis (i.e., with a very low risk of failure), while a short-term projection requires skills and the ability to act swiftly and adequately in often unpredictable and extraordinary situations. In both cases, the appropriate HCF is needed, and in both cases, the outcome depends to a great extent on the level of the MWL [6–8]. In the analysis that follows, we consider that the STA has its roots in the LTA and that the probability of the STA success depends to a great extent on the groundwork carried out when the LTA strategy and sequence of actions has been developed, as well as on the level and quality of training. "Hard in training, easy in battle," as the Russian General-in-Chief Suvorov put it. Better LTA facilitates STA. In other words, the STA has its roots in the LTA and can be viewed as a deviation from the LTA, when the aircraft is operated in conditions when STA is required. In this analysis, we assume that the probability of the STA success is distributed in accordance with the following double-exponential law of the extreme value distribution type [22]:

$$P^h(F,G) = P_0 \exp\left[\left(1 - \frac{G^2}{G_0^2}\right)\exp\left(1 - \frac{F^2}{F_0^2}\right)\right].$$ (8.13)

Here P_0 is the probability of success (nonfailure) of the LTA effort, which is characterized by the MWL for the specified (normal) LTA MWL level $G = G_0$, and the LTA HCF $F = F_0$; G_0 is the most likely (normal, specified, predetermined, and preestablished) LTA MWL; $G = G_0$ is the STA MWL; and $F \geq F_0$ is the required STA HCF. The P_0 level should be established beforehand, as a function of the G_0 level, when the HCF $F = F_0$. This could be done, for example, by conducting testing, measurements, and recordings on a flight simulator. The calculated ratios

$$\overline{P} = \frac{P^h(F,G)}{P_0} = \exp\left[\left(1 - \frac{G^2}{G_0^2}\right)\exp\left(1 - \frac{F^2}{F_0^2}\right)\right]$$ (8.14)

of the probability of the STA success to the probability of the LTA success are shown in Table 8.3. The following conclusions are drawn from the calculated data:

1. At normal MWL level ($G = G_0$) and/or at an extraordinarily (exceptionally) high HCF level ($F \to \infty$) the probability of the STA success is close to 100%.

2. The probabilities of the STA success are always lower than the probabilities of LTA success. This obvious fact is quantified by the calculated data.

3. If the MWL is exceptionally high, the STA effort will definitely fail, no matter how high his/her HCF is.

4. If the HCF is high, even a significant MWL has a small effect on the probability of the STA success, unless this workload is exceptionally large.

TABLE 8.3

Calculated $\bar{P} = P^h(F,G)/P_0$ Ratios of the Probability $P^h(F,G)$ of Human Nonfailure in Off-Normal Conditions to the Probability P_0 of Nonfailure in Normal Conditions

$\dfrac{G^2/G_0^2}{F^2/F_0^2}$	1	2	3	4	5	8	10	∞
1	1	0.3679	0.1353	0.0498	0.0183	9.1188E-4	1.234E-4	0
2	1	0.6922	0.4791	0.3317	0.2296	0.0761	0.0365	0
3	1	0.8734	0.7629	0.6663	0.5820	0.3878	0.2958	0
4	1	0.9514	0.9052	0.8613	0.8194	0.7057	0.6389	0
5	1	0.9819	0.9640	0.9465	0.9294	0.8797	0.8480	0
8	1	0.9991	0.9982	0.9978	0.9964	0.9936	0.9918	2.5E-40
10	1	0.9999	0.9998	0.9996	0.9995	0.9991	0.9989	4.4E-6
∞	1	1	1	1	1	1	1	1

5. The probability of STA success decreases with an increase in the MWL (especially for relatively low MWL levels) and increases with an increase in the HCF (especially for relatively low HCF levels). This intuitively obvious fact is quantified by the calculated data.

6. For high HCFs, the increase in the MWL level has a much smaller effect on the probabilities of STA success than for low HCFs.

All these conclusions make physical sense, of course, but provide a valuable quantitative assessment of the likelihood of the STA success.

Table 8.3 data show also that the increase in the F/F_0 ratio and in the G/G_0 ratio above the 3.0 value has a small effect on the probability of the STA success. This means particularly that an exceptionally highly qualified pilot does not have to be trained for an extraordinarily high STA-related MWL and does not have to be trained by a factor higher than 3.0 compared to a pilot of ordinary capacity (skills, qualification). In other words, a pilot does not have to be a superman to successfully cope with a high-level MWL in STA conditions, but still has to be trained in such a way that, when there is a need, he/she would be able to cope with a STA MWL by a factor of 3.0 higher than the normal level, and his/her STA HCF should be by a factor of 3.0 higher than what is expected of the same person in ordinary (normal) conditions.

From (8.14), we find, by differentiation,

$$\frac{d\bar{P}}{dG} = -\frac{2H}{G}\frac{G^2}{G^2 - G_0^2},$$

(8.15)

where $H = -\bar{P}\ln\bar{P}$ is the entropy (see, e.g., Ref. [17]) of the distribution of the relative probability of the STA success as compared to the LTA conditions. At the MWL levels close to the LTA level, the change in the relative probability of the STA success with the increase in the MWL level is significant. In another extreme case, when $G \gg G_0$, we have

$$\frac{d\bar{P}}{dG} = -\frac{2H}{G}. \qquad (8.16)$$

This formula explains the physical meaning of the distribution (8.13): the change in the probability of the STA success with the change in the level of the STA MWL is proportional, for large enough MWL levels, to the uncertainty level (entropy of the distribution of this probability) and is inversely proportional to the STA MWL level. The right part of the formula (8.16) could be viewed as a kind of a coefficient of variation, where the role of the uncertainty level in the numerator is played by the entropy, rather than by the standard deviation, and the role of the stress (loading) level in the denominator is played by the MWL level, rather than by the mean value of the random characteristic of interest.

8.5.2 LTA- and STA-Related MWL

Cognitive overload has been recognized as a significant cause of error in aviation, and therefore, measuring the MWL has become a key method of improving safety. There is an extensive published work in the psychological literature devoted to the measurement of MWL, both in military and in civil aviation. A pilot's MWL, including LTA and STA MWLs, can be measured using subjective ratings or objective measures. The subjective ratings during simulation tests can be in the form of periodic inputs to some kind of data collection device that prompts the pilot to enter a number between 1 and 10 (for example) to estimate the MWL every few minutes. Another possible approach is postflight paper questionnaires. There are some objective measures of MWL, such as heart rate variability. It is easier to measure the MWL on a flight simulator than in actual flight conditions. In a real airplane, one would probably be restricted to using postflight subjective (questionnaire) measures, since one would not want to interfere with the pilot's work.

An aircraft pilot faces numerous challenges imposed by the need to control a multivariate lagged system in a heterogeneous multitask environment. The time lags between critical variables require predictions and actions in an uncertain world. The interrelated concepts of situation awareness and MWL are central to aviation psychology. The major components of situation awareness are spatial awareness, system awareness, and task awareness. Each of these three components has real-world implications: spatial awareness for instrument displays, system awareness for keeping the operator informed about actions that have been taken by automated systems, and task awareness for attention and task management. Task management is directly related to the level of the MWL, as the competing "demands" of the tasks for attention might exceed the operator's resources—his/her "capacity" to adequately cope with the "demands" imposed by the MWL. In modern military aircraft, complexity of information, combined with time stress, creates difficulties for the pilot under combat conditions, and the first step to mitigate this problem is to measure and manage MWL. Although there is no universally accepted definition of the MWL and how it should/could be evaluated, there is a consensus that suggests that MWL can be conceptualized as the interaction between the structure of systems and tasks, on the one hand, and the capabilities, motivation, and state of the human operator, on the other. More specifically, MWL could be defined as the "cost" that an operator incurs as tasks are performed. Given the multidimensional nature of MWL, no single measurement technique can be expected to account for all the important aspects of it. Current research efforts in measuring MWL use psycho-physiological techniques, such as electroencephalographic, cardiac, ocular, and respiration measures in an attempt to identify and predict MWL levels. Measurement of cardiac activity has been the most popular physiological technique employed in the assessment of MWL, both from tonic variations in heart rate and after treatment of the cardiac signal. The authors of this paper intend to develop a methodology and to carry out experiments to measure the LTA and STA workloads.

8.5.3 LTA- and STA-Related HCF

The HCF includes the person's professional experience; qualifications; capabilities; skills; training; sustainability; ability to concentrate; ability to operate effectively, in a "tireless" fashion, under pressure, and, if needed, for a long period of time; ability to act as a "team player"; and swiftness of reaction (i.e., all the qualities that would enable him/her to cope with high MWL). In order to come up with a suitable figures-of-merit for the HCF, one could rank each of the above and other qualities on a scale from one to ten, and calculate the average FOM for each individual.

8.6 Conclusions

The following conclusions can be drawn from the carried-out analyses:

- There is a significant potential for reducing aeronautic casualties through better understanding the role that uncertainties of different natures play in the operator's world of work.

- We have shown how some of the uncertainties can be considered and quantified in two anticipation problems in aeronautics.

- The developed probabilistic models, with the appropriate modifications and generalizations, are applicable not only in the avionic and even not only in the vehicular domain, but in many situations, when a human encounters an uncertain and often harsh environment or any other hazardous situation, and/or interacts with never-perfect instrumentation and equipment, and when there is an incentive to quantify his/her qualifications and quantitatively assess his/her effort.

References

1. T.B. Sheridan and W.R. Ferrell, *Man-Machine Systems: Information, Control, and Decision Models of Human Performance*, MIT Press, Cambridge, MA, 1974.
2. M.R. Lehto and J.R. Buck, *Introduction to Human Factors and Ergonomics for Engineers*, Lawrence Erlbaum Associates, Taylor and Francis Group, New York and London, 2008.
3. D.C. Foyle and B.L. Hooey, *Human Performance Modeling in Aviation*, CRC Press, Boca Raton, FL, 2008.
4. E. Suhir, *Applied Probability for Engineers and Scientists*, McGraw-Hill, New York, 1997.
5. E. Suhir, "Helicopter-Landing-Ship: Undercarriage Strength and the Role of the Human Factor." In *ASME OMAE Conference*, Honolulu, Hawaii, June 1–9, 2009; see also *ASME OMAE Journal*, February 2010.
6. E. Suhir, "Probabilistic Modeling of the Role of the Human Factor in the Helicopter-Landing-Ship (HLS) Situation," *International Journal of Human Factor Modeling and Simulation (IJHFMS)*, vol. 1, No. 3, 2010, pp. 313–320.
7. E. Suhir and R.H. Mogford, "'Two Men in a Cockpit': Probabilistic Assessment of the Likelihood of a Casualty if One of the Two Navigators Becomes Incapacitated," *Journal of Aircraft*, vol. 48, No. 4, 2011, pp. 1309–1314.
8. E. Suhir, "'Human-in-the-Loop': Likelihood of Vehicular Mission-Success-and-Safety," *Journal of Aircraft*, vol. 49, No. 1, 2012, pp. 29–36.
9. E. Suhir, "'Miracle-on-the-Hudson': Quantitative Aftermath," *International Journal of Human Factor Modeling and Simulation*, vol. 4, No. 1, 2013, pp. 35–62.
10. A. Tversky and D. Kahneman, "Judgment under Uncertainty: Heuristics and Biases," *Science*, vol. 185, 1974, pp. 1124–1131.
11. D. Kahneman, P. Slovic, and A. Tversky (eds.), *Judgment under Uncertainty: Heuristics and Biases*, Cambridge University Press, New York, 1982.
12. L.P. Goodstein, H.B. Andersen, and S.E. Olsen (eds.), *Tasks, Errors, and Mental Models*, Taylor and Francis, Basingstoke, 1988.
13. S. Lini, C. Bey, S. Hourlier, B. Vallespir, A. Johnston, and P.-A. Favier, "Anticipation in Aeronautics: Exploring Pathways toward a Contextualized Aggregate Model Based on Existing Concepts." In D. de Waard, K. Brookhuis, F. Dehais, C. Weikert, S. Röttger, D. Manzey, S. Biede, F. Reuzeau, and P. Terrier (eds.), *Human Factors: A View from an Integrative Perspective*, 2012. *Proceedings HFES Europe Chapter Conference Toulouse*. ISBN 978-0-945289-44-9. Available from http://hfes-europe.org.

14. P. Denecker, "Les composantes symboliques et subsymboliques de l'anticipation dans lagestion des situations dynamiques," *Le Travail Humain*, vol. 62, No. 4, 1999, pp. 363–385.
15. A. Kaddoussi, N. Zoghlami, H. Zgaya, S. Hammadi, and F. Bretaudeau, "A Preventive Anticipation Model for Crisis Management Supply Chain." In *2011 IEEE International Conference on Systems, Man, and Cybernetics*, Anchorage, AK, October 9–20, 2011, pp. 109–114.
16. C. Mundutéguy and F. Darses, "Perception et anticipation du comportement d'autrui en situation simulée de conduite automobile," *Le Travail Humain*, vol. 70, No. 1, 2007, p. 32, Presses Universitaires de France, Paris.
17. R. Amalberti, "Anticipation." In M. de Montmollin (ed.), *Vocabulaire de l'Ergonomie*, Tolouse, Octares, 1995, pp. 42–43.
18. J.A. Michon, "The Complete Time Experience Time Experience." In J. A. Michon and J. L. Jackson (eds.), *Time, Mind and Behavior*, Heidelberger: Springer Verlag, 1985, pp. 20–52.
19. W. Friedman, "About Time: Inventing the Fourth Dimension," MIT Press, Cambridge, MA, August 1990, pp. 257–284.
20. J.M. Cellier, "Exigences et gestion temporelle dans les environnements dynamiques." In J.M. Cellier, V. De Keyser, and C. Valot (eds.), *La gestion du temps dans les environnements dynamiques*, Paris: Presses Universitaires de France, 1996, pp. 20–48.
21. J.M. Hoc, R. Amalberti, and G. Plee, "Vitesse du processus et temps partagé: planification et concurrence attentionnelle," *L'année psychologique*, vol. 100, No. 4, 2000, pp. 629–660.
22. E. Hollnagel, "Context, Cognition and Control." In Y. Waern (ed.), *Co-operative Process Management, Cognition and Information Technology*, 1998, pp. 27–52.
23. S. Lini, P.-A. Favier, J.-M. André, S. Hourlier, C. Bey, B. Vallespir, and B. Baracat, "Influence of Anticipatory Time Depth on Cognitive Load in an Aeronautical Context," *Le Travail Humain*, vol. 78, No. 3, 2012, pp. 193–215, Presses Universitaires de France, Paris.
24. P.A. Hancock and J.K. Caird, "Experimental Evaluation of a Model of Mental Workload," *Human Factors: Journal of the Human Factors and Ergonomics Society*, vol. 35, No. 3, 1993, pp. 413–429.

9

Double-Exponential Probability Distribution Function for the Human Nonfailure

9.1 Summary

The probabilistic predictive modeling (PPM) approach in human-in-the-loop (HITL) related aerospace problems enables one to predict, quantify, assure, and even specify the probability of the outcome of an aerospace mission or a situation, when the performance of a never-perfect human, never-100%-reliable instrumentation (equipment), never absolutely predictable response of the object of control (aero- or spacecraft, or even an ocean-going vessel), uncertain and often harsh environment, as well as the interaction (interfaces) of these uncertainties, contribute jointly to the likelihood of such an outcome. While the reliability of the navigation instrumentation (equipment) could be evaluated using the well-known suitable and more or less well-established modeling means, such as, e.g., probabilistic design for reliability (PDfR), the role of the human factor (HF) could be considered, when quantification of the role of the HITL is critical, by using, as has been shown in Chapters 6, 9, and 10, the double-exponential probability distribution function (DEPDF). It is noteworthy that our DEPDF is not one of the classical extreme value distributions (EVDs), which are also double-exponential ones.

In the recently published paper [1], it was suggested to account, in addition to the mental workload (MWL) and human capacity factor (HCF), also for the possible role of the current (measured/monitored) human's state of health (SH) that could have an appreciable effect on the perceived MWL:

$$P^h\left(F,G,S_*\right) = P_0\ \exp\left[\left(1-\gamma_S S_* t-\frac{G^2}{G_0^2}\right)\exp\left(1-\frac{F^2}{F_0^2}\right)\right]. \tag{9.1}$$

In the analysis that follows, we consider, in addition to MWL, HCF, and SH, also the factor HE T_* of the propensity of a human to make an error:

$$P^h\left(F,G,S_*\right) = P_0 \exp\left[\left(1-\gamma_S S_* t-\frac{G^2}{G_0^2}\right)\exp\left(1-\gamma_T T_* -\frac{F^2}{F_0^2}\right)\right] \tag{9.2}$$

(See all the notation taken in this equation in Section 9.3.) This has been done assuming and suggesting that mean time to failure (MTTF) T_* of a human, performing his/her professional duties, can be used as an adequate criterion of his/her failure-free performance: when this performance is error free, the MTTF is infinitely long and is very short in an opposite extreme case. The suggested DEPDF considers that both high MTTF and high HCF result in a higher probability of a failure-free human performance, but, unlike in the previously suggested DEPDF structure, enables one to separate the MTTF as the direct HF characteristic from other critical HCF features, such as, for example, level of training, ability to operate under time pressure, and mature thinking. The suggested SH characteristic is separated in the suggested DEPDF formulation from the general MWL level to emphasize the importance of human SH that could affect his/her perception of the otherwise more or less objective MWL. In any event, the major goal of this chapter is to generate thinking on how to advance the state of the art in today's aerospace-related human psychology, and, particularly, on how to quantify, by both modeling and experimentation, the HITL-related effort, when the HF and equipment/instrumentation performance contribute jointly to the success and safety of an aerospace mission or an extraordinary situation. Future work will certainly be aimed at these efforts.

9.2 Introduction

The subject of this chapter can be defined as probabilistic ergonomics, probabilistic human factor (HF) engineering, or probabilistic manned-systems technology. The chapter is geared primarily to the HITL-related situations, when human performance and equipment reliability contribute jointly to the outcome of a mission or a situation, and since nobody and nothing is perfect, this should be done on the probabilistic basis.

While considerable improvements in various aerospace missions and extraordinary situations can be achieved through better traditional ergonomics, better work environment, and other well-established human psychology means that affect directly the individual's behavior and performance, there is also a significant potential for improving safety in the air and in the outer space by quantifying the HF. Since nobody is perfect, such a quantification should be done by using PPM and probabilistic risk analysis (PRA) methods and approaches (see, e.g., Refs. [2,3]). The rationale behind such intent and the incentive can be explained by the following more or less obvious reasoning:

- When the outcome of an aerospace mission or a situation is imperative, and this is typically the case in the majority of the HITL-related situations, the ability to predict and to quantify such an outcome

Double-Exponential Probability Distribution 197

is imperative, otherwise, the success and safety of the aerospace mission of importance or the favorable outcome of an anticipated off-normal situation simply cannot be assured.

- The outcome of an aerospace mission or a situation is affected by the combined action of various human-performance factors, instrumentation and equipment reliability, response of the object of control (the air- or the spacecraft) to the navigator's actions, uncertain and often harsh environment, as well as by the interaction of these factors, so that quantification of the role of various causes affecting the outcome of an aerospace mission or a situation should be properly considered in order to make a logical, physically meaningful, timely, and practically useful prediction of the outcome and to most effectively handle its consequences.

- To be practical, such a prediction should be based on (geared to) an adequate, simple, easy-to-use, physically meaningful, and logically justifiable predictive model that reflects the most important features of the mission or the situation of interest.

- Since nobody and nothing is perfect, such a model should use applied probability and PRA methods and approaches; it is clear that the difference between a highly reliable system, a failure-free human performance, or a successful outcome of a mission or a situation or of insufficiently reliable ones is "merely" in the level of their never-zero probability of failure.

- In the recently suggested probabilistic design for reliability (PDfR) approach, whether in application to the equipment and instrumentation or the HITL, the planning effort should be geared to a prior probabilistic, predictive model that adequately reflects the main features and important attributes of the phenomenon or mission of interest, rather than by relying on a posterior, statistical, strategy; such a model should be, of course, as simple as possible, but be physically meaningful and provide practically useful information.

- When the physical substance and attributes of an anticipated mission or a situation of interest could be described by one of the known probabilistic distributions (normal, log-normal, Rayleigh, Weibull, etc.), then it is this distribution that should be employed in a particular HITL analysis, and the probabilistic characteristics of such a distribution (mean value, standard deviation, etc.) should be based on and substantiated by the conducted and trustworthy statistics.

- While the MWL level is always important and should be always considered when addressing and evaluating an outcome of a mission or a situation, the HCF is equally important: the same MWL can result in a completely different outcome depending on the HCF level of the individual(s) involved; in other words, it is the relative levels of the

198   *Human-in-the-Loop*

MWL and HCF that have to be considered in one way or another, when assessing the likelihood of a mission or a situation success and safety.

- MWL and HCF can be characterized by different means and different measures, but it is clear that both of these factors have to have the same units in a particular problem of interest. It should be emphasized that one important and favorable consequence of an effort based on the consideration of the above factors is bridging the existing gap between what the aerospace psychologists and system analysts do. Based on the author's interactions with aerospace system analysts and avionic human psychologists, these two categories of specialists seldom team up and collaborate. Application of the PPM/PRA concept provides, in the author's opinion, a natural and an effective means for quantifying the expected HITL-related outcome of a mission or a situation and for minimizing the likelihood of a mishap, casualty, or failure [4–8]. By employing quantifiable and measurable ways of assessing the role and significance of various uncertainties and by treating HITL-related missions and situations as part, often the most crucial part, of the complex man-instrumentation-equipment-vehicle-environment system, one could improve dramatically the human performance and the state of the art in assuring aerospace missions success and safety [7,9,10].

The PPM/PRA concept enables one to predict, quantify, assure, and, if needed, even specify the adequate and never-zero probability of the occurrence of a mishap. This probability cannot be high, of course, but does not have to be lower than necessary either: to be economically feasible it has to be adequate for a particular application, mission, or situation [11]. The best engineering design is, in effect, the best compromise between different and comprehensive requirements. If one intends to optimize reliability with an objective to achieve the most feasible compromise between the adequate reliability, minimum cost, and minimum time for the completion of the project ("time-to-market"), ability to quantify reliability versus cost and planning time is imperative. There is always an incentive, therefore, for optimizing equipment and human performance in terms of cost, preparation (planning), and training times, as well as the availability and effectiveness of the anticipated effort [12]. No optimization is possible, if the instrumentation reliability and human performance characteristics of interest are not quantified, and if the underlying physics of the equipment/instrumentation failure and the pertinent human psychology factors are not well understood, effectively predicted, and brought to an acceptable level. Of course, at today's state of the art in both the operational reliability of the equipment/instrumentation and human performance, such a quantification is not easy to accomplish, but one should have in mind that such an objective should be pursued to an extent possible.

The important feature of our approach is that it starts with an adequate predictive model. While the traditional statistical (posterior) HITL-oriented

Double-Exponential Probability Distribution

approaches are based on experimentations followed by statistical analyses, our PPM/PRA (*a priori*) concept is based on, and starts with, physically meaningful and flexible predictive modeling followed, if necessary, by highly focused and highly cost-effective experimentations. Such experimentations are mostly of the FOAT type: testing is conducted until failure, whatever its definition might be. Such a FOAT is geared to the particular chosen simple, easy-to-use, more or less well-established and physically meaningful governing model [13]. It is expected that appropriate practically useful figures-of-merit (FOM) could be eventually developed as a result of such an effort. If the predicted probability of an outcome of a mission or an off-normal situation is not acceptable, then an appropriate sensitivity analysis (SA) based on the developed and available PPM/PRA methodologies and algorithms can be effectively employed to improve the situation.

The following major results were obtained on the role of the human factor with an emphasis on modeling efforts. General aspects of the role of the human factor were addressed by Goodstein, Andersen, and Olsen [14]; Reason [15]; Hollnagel [16,17]; Archer and Allender [18]; Jagacinski and Flach [19]; Gluck and Pew [20]; and Lehto and Buck [21]. Various human performance models in aviation, with an emphasis on the aviation safety were developed by Orasanu, Martin, and Davison [22]; Leiden, Keller, and French [23]; Leveson [24]; Hobbs [25]; Foyle and Hooey [26]; and Hourlier and Suhir [8]. Long- and short-term preventive anticipation in aviation has been modeled and analyzed by Amalberti [27]; Denecker [28]; Hoc, Amalberti, and Plee [29]; Mundutéguy and Darses [30]; Kaddoussi, Zoghlami, Zgaya, Hammadi, and Bretaudeau [31]; Lini, Bey, Hourlier, Vallespir, Johnston, and Favier [32]; and Lini, Favier, André, Hourlier, Bey, Vallespir, and Baracat [33]. Probabilistic assessments were carried out by Suhir, Bey, Lini, Salotti, Hourlier, and Claverie [34]. The importance of quantitative modeling has been indicated first by Restle and Greeno [35], and later on addressed by Baron, Kruser, and Huey [36]; Polkand Seifert [37]; Wickens [38]; Estes [39]; and Sargent [40]. Probabilistic approaches were employed in the HITL problems for about 50 years or so to account for various uncertainties in the problems in question. Here are several publications in this field: Tversky and Kahneman [41]; Kahneman, Slovic, and Tversky [42]; Rasmussen and Pedersen [43]; Gore and Smith [44]; Suhir [10,45–47]; and Suhir and Mogford [48]. Various aspects of the role of the human factor in interplanetary missions were addressed and analyzed by Salotti and Claverie [49]; Salotti [50]; Salotti and Suhir [51,52]; and Salotti, Hedmann, and Suhir [53]. The PPM/PRA approach considered in this chapter is based, like in Ref. [1], on the application of the DEPDF. The difference is that in Ref. [1] the role of the HE aspect was considered indirectly, through the level and role of the HCF.

In this analysis, however, the HE characteristic is introduced directly, assuming that the MTTF of a human, when performing his/her duties, is an adequate criterion of his/her failure-free performance: in the case of an error-free performance this time is infinitely long, and is short in an opposite case.

The suggested expression for the DEPDF considers that both high MTTF and high HCF result in a higher probability of a failure-free human performance, but enables one to separate the MTTF as the direct HF characteristic from other HCF features, such as, for example, level of training, ability to operate under time pressure, and mature thinking. The DEPDF could be introduced, as has been shown in the author's previous publications, in many different ways depending on the particular mission or a situation, and on the sought information. The DEPDF suggested in this analysis considers the following major factors: flight duration, the acceptable level of the continuously monitored (measured) human SH characteristic (symptom), the MTTF as an appropriate HE characteristic, the level of the MWL, and the HCF. While the notion of the MWL is well established in aerospace psychology and is reasonably well understood and investigated, the notion of the HCF was introduced by the author of this analysis only several years ago. The rationale behind that notion is that it is not the absolute MWL level, but the relative levels of the MWL and HCF that determine, in addition to other critical factors, the probability of the human failure. It has been shown [1] that the DEPDF has its physical roots in the entropy of this function. It is shown also how the DEPDF could be established from the highly focused and highly cost-effective FOAT data. FOAT is a must, if understanding the physics of failure of instrumentation and/or of human performance is imperative to assure high likelihood of a failure-free aerospace operation. The FOAT data could be obtained by testing on a flight simulator, by analyzing the responses to postflight questionnaires, or by using Delphi technique. FOAT could not be conducted, of course, in application to human health, but testing and SH monitoring could be run until a certain level (threshold) of the human SH characteristic (symptom), still harmless to his/her health, is reached. The general concepts addressed in our analysis are illustrated by practical numerical examples. It is demonstrated how the probability of a successful outcome of the anticipated aerospace mission can be assessed in advance, prior to the fulfillment of the actual operation. Although the input data in these examples are more or less hypothetical, they are, nonetheless, realistic. These examples should be viewed, therefore, as useful illustrations of how the suggested DEPDF model can be implemented. It is the author's belief that the developed methodologies, with appropriate modifications and extensions, when necessary, can be effectively used to quantify, on the probabilistic basis, the roles of various critical uncertainties affecting success and safety of an aerospace mission or a situation of importance. The author believes also that these methodologies and formalisms can be used in other cases, well beyond the aerospace domain, when a human encounters an uncertain environment or an hazardous off-normal situation, and when there is an incentive/need to quantify his/her qualifications and performance, and/or when there is a need to assess and possibly improve the human role in a particular mission or a situation, and/or when there is an intent to include this role into an analysis of interest, with consideration of the

Double-Exponential Probability Distribution 201

navigator's SH. Such an incentive always exists for astronauts in their long outer space journeys, but could be also of importance for long enough aircraft flights, when, for example, one of the two pilots gets incapacitated during the flight [48]. The PPM/PRA concept, including the use of the DEPDF, has been applied during the last several years to some tasks and situations in the aerospace vehicular engineering, namely, (1) to evaluate the helicopter undercarriage strength when landing on a ship deck, preferably during a short-term lull (calm window) in the wave conditions, when the random times required for the officer on ship board and the helicopter pilot to make their "go–no go" decisions and the time of actual landing have to be considered against the random lull time to judge on the likelihood of safe landing [4,6,9,45,46]; (2) to predict the likelihood of an aircraft mission success by considering human–instrumentation interactions at different segments of the route [4,5,10,46]; (3) to explain that the actual "miracle" in the famous "Miracle-on-the-Hudson" ditching should be attributed, first, to Captain Sullenberger's exceptionally high HCF that enabled him to successfully cope with the extraordinarily high MWL [6]; although the HCFs of Captain Sullenberger and of the "UN-shuttle" crew were not actually measured, this chapter provides, nonetheless, a flavor of the type of HCF-related qualities that supposedly have played a crucial role in the two events addressed, and some of these qualities (see Section 9.3) could have been established beforehand, for example, by testing pilots on flight simulators and/or by reviewing and analyzing postflight questionnaires and/or by applying Delphi method, and/or by using any other suitable means; (4) to address several anticipation-related problems in avionics, such as, for example, short- and long-term anticipations [34]; (5) to assess the likelihood of a casualty if one of the two navigators becomes incapacitated during the flight [48]; and (6) to address and discuss some important attributes of, and challenges in, applying PPM/PRA techniques in various HITL problems [7,9], including those associated with the future manned missions to Mars [49,51–53]. The PPM/PRA treatment of the HITL-related problems was viewed in these applications as part of a more general PDfR concept aimed at the prediction of the possible failures in aerospace electronic and photonic engineering, with consideration of the human factor (see, e.g., Refs. [2,3,46,47,54–58]).

The analysis carried out in this chapter is, in effect, an extension of the above effort and is focusedon the application of the DEPDF in those HITL-related problems in aerospace engineering that are aimed at the quantification, on the probabilistic basis, of the role of the HF, when both the human performance and, particularly, his/her SH affect the success and safety of an aerospace mission or a situation. While the PPM of the reliability of the navigation instrumentation (equipment), both hard- and software, could be carried out using well-known Weibull distribution or on the basis of the recently suggested Boltzmann–Arrhenius–Zhurkov equation (see, e.g., Ref. [58]), or other suitable and more or less well-established means, the role of the human factor, when quantification of the human role is critical, could be considered by using the suggested DEPDF. There might be other ways to go, but this is, in the author's view and experience, a quite natural and a rather effective way.

The DEPDF is of the EVD type (see, e.g., Refs. [2,59,60])—that is, places an emphasis on the inputs of extreme loading conditions that occur in extraordinary (off-normal) situations. The DEPDF reflects, however, a probabilistic *a priori*–type, rather than a statistical *a posteriori*–type approach, and could be introduced in many ways (see, e.g., Refs. [2,4,5,6,9,48,51]), depending on the particular mission or a situation, as well as on the sought information. It is noteworthy that the DEPDF is not a special case, nor a generalization, of [59,60], or any other well-known statistical EVD used for many decades in various applications of the statistics of extremes, such as, for example, prediction of the likelihood of extreme earthquakes or floods. Our DEPDF should be rather viewed as a practically useful aerospace engineering–related relationship that makes physical and logical sense in many practical problems and situations, and could and should be employed when there is a need to quantify the probability of the successful outcome of a HITL-related aerospace mission or a situation. The DEPDF suggested in this analysis considers the following major factors: the flight/operation duration; the acceptable level of the continuously monitored (measured) meaningful human SH characteristic (FOAT approach is not acceptable in this case); the MWL level; the MTTF as an appropriate HE characteristic; and the HCF. While the notion of the MWL is well established in aerospace psychology and is reasonably well understood and investigated, the notion of the HCF was introduced by the author only several years ago. The rationale behind such an introduction is that it is not the absolute MWL level, but the relative levels of the MWL and HCF that determine, in addition to other critical factors, the probability of the human nonfailure in a particular off-normal situation of interest. The majority of pilots with an ordinary HCF would fail in the "Miracle-on-the-Hudson" situation, while "Sully," with his extraordinarily high anticipated HCF [6] did not. As to the quantified HE characteristic, the MTTF is introduced in this chapter as such a characteristic for the first time. Various aspects of SH and HE characteristics are intended to be addressed in the author's future work as important items of an outer space medicine. The recently suggested three-step concept methodology [54] is intended to be also employed in such an effort.

9.3 DEPDF with Consideration of Time, Human Error, and State of Health

The DEPDF could be introduced, as has been mentioned, in many ways, and its particular formulation depends on the problem addressed. In this

Double-Exponential Probability Distribution

analysis we suggest a DEPDF that enables one to evaluate the impact of three major factors, the MWL G, the HCF F, and the time t (possibly affecting the navigator's performance and sometimes even his/her health), on the probability $P^h(F,G,t)$ of his/her nonfailure. While measuring the MWL has become a key method of improving aviation safety (see Chapter 4), the HCF has been introduced by the author just several years ago as an important characteristic to be considered when assessing the outcome of a particular mission or a situation of interest. With an objective to quantify the likelihood of the human nonfailure, the corresponding probability is sought in the form of the following DEPDF:

$$P^h\left(F,G,S_*\right) = P_0 \exp\left[\left(1 - \gamma_S S_* t - \frac{G^2}{G_0^2}\right)\exp\left(1 - \gamma_T T_* - \frac{F^2}{F_0^2}\right)\right]. \qquad (9.1)$$

Here P_0 is the probability of the human nonfailure at the initial moment of time ($t = 0$) and at a normal (low) level of the MWL ($G = G_0$), S_* is the threshold (acceptable level) of the continuously monitored/measured (and possibly cumulative, effective, indicative, even multiparametric) human health characteristic (symptom), such as, for example, body temperature, arterial blood pressure, oxy-hemo-metric determination of the level of saturation of blood hemoglobin with oxygen, electrocardiogram measurements, pulse frequency and fullness, frequency of respiration, measurement of skin resistance that reflects skin covering with sweat, and so on (since the time t and the threshold S_* enter the expression [9.1] as a product $S_* t$ each of these parameters has a similar effect on the sought probability [9.1]); γ_S is the sensitivity factor for the symptom S_*; $G \geq G_0$ is the actual (elevated, off-normal, extraordinary) MWL that could be time dependent; G_0 is the MWL in ordinary (normal) operation conditions; T_* is the MTTF; γ_T is the sensitivity factor for the MTTF T_*; $F \geq F_0$ is the actual (off-normal) HCF exhibited or required in an extraordinary condition of importance; and F_0 is the most likely (normal, specified, ordinary) HCF. It is clear that there is a certain overlap between the levels of the HCF F and the T_* value, which has also to do with the human quality. The difference is that the T_* value is a short-term characteristic of the human performance that might be affected, first, by his/her personality, while the HCF is a long-term characteristic of the human, such as his/her education, age, experience, and ability to think and act independently. The author believes that the MTTF T_* might be determined for the given individual during testing on a flight simulator, while the factor F, although should be also quantified, cannot be typically evaluated experimentally, using accelerated testing on a flight simulator. While the P_0 value is defined as the probability of nonfailure at a very low level of the MWL G it could be determined and evaluated also as the probability of nonfailure for a hypothetical situation when the HCF F is extraordinarily high (i.e., for an individual/pilot/navigator who is exceptionally highly qualified), while the MWL G is still finite, and so is the operational time t.

Note that the function (9.2), unlike the DEPDF suggested in Ref. [1], has a nice symmetric and consistent form. It reflects, in effect, the roles of the MWL + SH "objective," "external" impact $E = \left(1 - \gamma_S S_* t - \dfrac{G^2}{G_0^2}\right)$, and of the HCF + HE "subjective," "internal" impact $I = \left(1 - \gamma_T T_* - \dfrac{F^2}{F_0^2}\right)$. The rationale below the structures of these expressions is that the level of the MWL could be affected by the human's SH (the same person might experience a higher MWL, which is not only different for different humans, but might be quite different depending on the navigator's SH), while the HCF, although could also be affected by his/her SH, has its direct measure in the likelihood that he/she makes an error. In our approach this circumstance is considered by the T_* value, MTTF, since an error is, in effect, the failure to an error-free performance. When the human's qualification is high, the likelihood of an error is lower. The "external" $E = $ MWL + SH factor is more or less a short-term characteristic of the human performance, while the factor $I = $ HCF + HE is a more permanent, more long-term characteristic of the HCF and its role.

9.4 MWL

Measuring the MWL has become a key method of improving aviation safety, and there is an extensive published work devoted to the measurement of the MWL in aviation, both military and commercial. A pilot's MWL can be measured using subjective ratings and/or objective measures. The subjective ratings during FOAT (simulation tests) can be, for example, after the expected failure is defined, in the form of periodic inputs to some kind of data collection device that prompts the pilot to enter, for example, a number between 1 and 10 to estimate the MWL every few minutes.

There are also some objective MWL measures, such as, for example, heart rate variability. Another possible approach uses postflight questionnaire data: it is usually easier to measure the MWL on a flight simulator than in actual flight conditions. In a real aircraft, one would probably be restricted to using postflight subjective (questionnaire) measurements, since a human psychologist would not want to interfere with the pilot's work. Given the multidimensional nature of MWL, no single measurement technique can be expected to account for all the important aspects of it. In modern military aircraft, complexity of information, combined with time stress, creates significant difficulties for the pilot under combat conditions, and the first step to mitigate this problem is to measure and manage the MWL.

Current research efforts in measuring MWL use psycho-physiological techniques, such as electroencephalographic, cardiac, ocular, and respiration measures in an attempt to identify and predict MWL levels. Measurement of cardiac activity has been also a useful physiological technique employed in the assessment of MWL, both from tonic variations in heart rate and after treatment of the cardiac signal. Such an effort belongs to the fields of astronautic medicine and aerospace human psychology. Various aspects of the MWL, including modeling, and situation awareness analysis and measurements, were addressed by Bundesen [61]; Hamilton and Bierbaum [62]; Baddeley [63]; Hancock and Caird [64]; Endsley [65]; Ericsson and Kintsch [66]; Cellier [67]; Endsley and Garland [68]; Lebiere [69]; Kirlik [70]; and Diller, Gluck, Tenney, and Godfrey [71].

9.5 HCF

HCF, unlike MWL, is a new notion [45]. Taatgen [72] was probably the first one who developed a model to account for individual differences in application to learning air traffic control. Among other publications on quantitative assessments of the role of the human factor we would like to indicate those of Baron, Kruser, and Huey [36]; Card, Moran, and Newell [73]; Friedman [74]. HCF plays with respect to the MWL approximately the same role as strength/capacity plays with respect to stress/demand in structural analysis and in some economics problems. HCF includes, but might not be limited to, the following major qualities that would enable a professional human to successfully cope with an elevated off-normal MWL (see, e.g., Ref. [6]): psychological suitability for a particular task; professional experience and qualifications; education, both special and general; relevant capabilities and skills; level, quality, and timeliness of training; performance sustainability (consistency, predictability); independent thinking and independent acting, when necessary; ability to concentrate; ability to anticipate; self-control and ability to act in cold blood in hazardous and even life-threatening situations; mature (realistic) thinking; ability to operate effectively under pressure, and particularly under time pressure; leadership ability; ability to operate effectively, when necessary, in a tireless fashion, for a long period of time (tolerance to stress); ability to act effectively under time pressure and make well-substantiated decisions in a short period of time and in uncertain environmental conditions; team-player attitude, when necessary; and swiftness in reaction, when necessary. These and other qualities are certainly of different importance in different HITL situations. It is clear also that different individuals possess these qualities in different degrees. Long-term HCF could be time dependent. To come up with suitable FOM for the HCF, one could rank, similarly to the MWL estimates, the above and perhaps other qualities on the scale from, say, 1 to 10, and calculate the average FOM for each individual and particular task. Clearly, MWL and HCF should use the same measurement units, which could be particularly nondimensional. Special psychological tests might be necessary to develop and conduct to establish the level of these qualities for the individuals of significance.

9.6 The Introduced DEPDF Makes Physical Sense

The function (9.1) makes physical sense. Indeed,

1. When time t, and/or the level S_* of the governing SH symptom, and/ or the level of the MWL G are significant, the probability of non-failure is always low, no matter how high the level of the HCF F might be.
2. When the level of the HCF F and/or the MTTF T_* are significant, and the time t, and/or the level S_* of the governing SH symptom, and/or the level of the MWL G are finite, the probability $P^h(F,G,S_*)$ of the human nonfailure becomes close to the probability P_0 of the human nonfailure at the initial moment of time $(t = 0)$ and at a normal (low) level of the MWL $(G = G_0)$.
3. When the HCF F is on the ordinary level F_0 the formula (9.1) yields

$$P^h(F,G,S_*) = P^h(G,S_*) = P_0 \exp\left[\left(1 - \gamma_S S_* t - \frac{G^2}{G_0^2}\right) \exp(-\gamma_T T_*)\right]. \quad (9.2)$$

For a long time in operation $(t \to \infty)$ and/or when the level S_* of the governing SH symptom is significant $(S_* \to \infty)$ and/or when the level G of the MWL is high, the probability of nonfailure will always be low, provided that the MTTF T_* is finite.

4. At the initial moment of time $(t = 0)$ and/or for the very low level of the SH symptom $S_*(S_* = 0)$, the formula (9.1) yields

$$P^h(F,G,T_*) = P^h(G) = P_0 \exp\left[\left(1 - \frac{G^2}{G_0^2}\right) \exp\left(1 - \gamma_T T_* - \frac{F^2}{F_0^2}\right)\right]. \quad (9.3)$$

When the MWL G is high, the probability of nonfailure is low, provided that the MTTF T_* and the HCF F are finite. However, when the HCF is extraordinarily high and/or the MTTF T_* is significant (low likelihood that HE will take place), the probability (9.3) of nonfailure will be close to one. In connection with the taken approach and particularly with the basic equation (9.1), it is noteworthy also that not every model needs prior experimental validation. In the author's view, the structure of the model (9.1) does not. Just the opposite: this model should be used as the basis of FOAT-oriented accelerated experiments to establish the MWL, the HCF, and the levels of

Double-Exponential Probability Distribution 207

HE (through the corresponding MTTF) and his/her SH at normal operation conditions and for a navigator with regular skills and of ordinary capacity. These experiments could be run, for example, on different flight simulators and on the basis of specially developed testing methodologies. Being a probabilistic, not a statistical model, Equation (9.1) should be used to obtain, interpret, and accumulate relevant statistical information. Starting with collecting statistics first seems to be a time-consuming and highly expensive path to nowhere.

9.7 Underlying Reliability Physics

Assuming, for the sake of simplicity, that the probability P_0 is established and differentiating the expression

$$\bar{P} = \frac{P^h(F,G,S_*)}{P_0} = \exp\left[\left(1-\gamma_S S_* t - \frac{G^2}{G_0^2}\right)\exp\left(1-\gamma_T T_* - \frac{F^2}{F_0^2}\right)\right] \tag{9.4}$$

with respect to the time t the following formula can be obtained:

$$\frac{d\bar{P}}{dt} = -H(\bar{P})\frac{1-\gamma_S S_* - \dfrac{G^2}{G_0^2}}{1-\gamma_S S_* t - \dfrac{G^2}{G_0^2}}, \tag{9.5}$$

where $H(\bar{P}) = -\bar{P}\ln\bar{P}$ is the entropy of the distribution (9.4). When the MWL G is on its normal level G_0 and/or when the still accepted SH level S_* is extraordinarily high, the formula (9.5) yields

$$\frac{d\bar{P}}{dt} = -\frac{h(\bar{P})}{t}. \tag{9.6}$$

Hence, the basic distribution (9.1) is a generalization of the situation, when the decrease in the probability of human performance nonfailure with time can be evaluated as the ratio of the entropy $H(\bar{P})$ of the distribution (9.4) to the elapsed time t, provided that the MWL is on its normal level and/or the HCF of the navigator is exceptionally high.

At the initial moment of time ($t = 0$) and/or when the governing symptom has not yet manifested itself ($S_* = 0$), the distribution (9.4) yields

$$\bar{P} = \exp\left[\left(1-\frac{G^2}{G_0^2}\right)\exp\left(1-\gamma_T T_* - \frac{F^2}{F_0^2}\right)\right] \tag{9.7}$$

Then we find,

$$\frac{d\overline{P}}{dG} = 2H(\overline{P})\frac{\dfrac{G^2}{G_0^2}}{1-\dfrac{G^2}{G_0^2}}. \tag{9.8}$$

For significant MWL levels, this formula yields

$$\frac{d\overline{P}}{dG} = -2H(\overline{P}). \tag{9.9}$$

Thus, another way to interpret the underlying physics of the distribution (9.1) is to view this distribution as such that considers that the change in the probability (9.4) at the initial moment of time with the change in the level of the MWL and when this level is significant, is twice as high as the entropy of the distribution (9.4).

The entropy $H(\overline{P})$ is zero for the probabilities $\overline{P} = 0$ and $\overline{P} = 1$, and reaches its maximum value $H_{max} = \dfrac{1}{e} = 0.3679$ for $\overline{P} = \dfrac{1}{e} = 0.3679$. Hence, the derivative $\dfrac{d\overline{P}}{dG}$ is zero for the probabilities $\overline{P} = 0$ and $\overline{P} = 1$, and its maximum value $\left(\dfrac{d\overline{P}}{dG}\right)_{max} = \dfrac{2}{eG} = \dfrac{0.7358}{G}$ takes place for $\overline{P} = \dfrac{1}{e} = 0.3679$.

The $\overline{P}$ values calculated for the case $T_* = 0$ (human error is likely, but could be rapidly corrected because of the high HCF) indicate that

1. At normal MWL level and/or at an extraordinarily (exceptionally) high HCF level the probability of human nonfailure is close to 100%.

2. If the MWL is exceptionally high, the human will definitely fail, no matter how high his/her HCF is.

3. If the HCF is high, even a significant MWL has a small effect on the probability of nonfailure, unless this MWL is exceptionally large (indeed, highly qualified individuals are able to cope better with various off-normal situations and get tired less when time progresses than individuals of ordinary capacity).

4. The probability of nonfailure decreases with an increase in the MWL (especially for relatively low MWL levels) and increases with an increase in the HCF (especially for relatively low HCF levels).

5. For high HCFs the increase in the MWL level has a much smaller effect on the probabilities of nonfailure than for low HCFs; it is noteworthy that the above intuitively more or less obvious judgments can be effectively quantified by using analyses based on Equations (9.1) and (9.4).

Double-Exponential Probability Distribution 209

6. The increases in the HCF (F/F_0 ratio) and in the MWL (G/G_0 ratio) above the 3.0 have a minor effect on the probability of nonfailure; this means particularly that the navigator does not have to be trained for an extraordinarily high MWL and/or possess an exceptionally high HCF (F/F_0 ratio), higher than 3.0, compared to a navigator of an ordinary capacity (qualification); in other words, a navigator does not have to be a superman or a superwoman to successfully cope with a high-level MWL, but still has to be trained to be able to cope with a MWL by a factor of three higher than the normal level.

If the requirements for a particular level of safety are above the HCF for a well-educated and well-trained human, then the development and employment of the advanced equipment and instrumentation should be considered for a particular task, and the decision about the right way to go should be based on the evaluation, on the probabilistic basis, of both the human and the equipment performance, and the possible consequences of failure.

9.8 Possible FOAT-Based Procedure to Establish the DEPDF

In the basic DEPDF (1) there are three unknowns: the probability P_0 and two sensitivity factors γ_S and γ_T. As mentioned previously, the probability P_0 could be determined by testing the responses of a group of exceptionally highly qualified individuals, such as, for example, Captain Sullenberger in the famous Miracle-on-the-Hudson event.

Let us show how the sensitivity factors γ_S and γ_T can be determined. Equation (9.4) can be written as

$$\frac{\ln \bar{P}}{1 - \gamma_S S_* t - \dfrac{G^2}{G_0^2}} = \exp\left(1 - \gamma_T T_* - \frac{F^2}{F_0^2}\right). \tag{9.10}$$

Let FOAT be conducted on a flight simulator for the same group of individuals, characterized by the more or less the same high MTTF T_* values and high HCF $\dfrac{F}{F_0}$ ratios, at two different elevated (off-normal) MWL conditions, G_1 and G_2. The governing symptom has reached its critical preestablished level S_* at the times t_1 and t_2 from the beginning of testing, respectively, and the corresponding percentages of the individuals that failed the tests were Q_1 and Q_2, so that the corresponding probabilities of nonfailure were $\bar{P}_1$ and $\bar{P}_2$, respectively. Since the same group of individuals was tested, the right part of Equation (9.10) that reflects the levels of the

HCF and HE remains more or less unchanged, and therefore, the following requirement should befulfilled:

$$\frac{\ln \bar{P}_1}{1 - \gamma_S S_* t_1 - \dfrac{G_1^2}{G_0^2}} = \frac{\ln \bar{P}_2}{1 - \gamma_S S_* t_2 - \dfrac{G_2^2}{G_0^2}}. \tag{9.11}$$

This equation yields

$$\gamma_S = \frac{1}{S_*} \frac{1 - \dfrac{G_1^2}{G_0^2} - \dfrac{\ln \bar{P}_1}{\ln \bar{P}_2}\left(1 - \dfrac{G_2^2}{G_0^2}\right)}{t_1 - \dfrac{\ln \bar{P}_1}{\ln \bar{P}_2} t_2}. \tag{9.12}$$

After the sensitivity factor γ_S for the assumed symptom level S_* is determined, the dimensionless variable $\gamma_T T_*$, associated with the human error sensitivity factor γ_T could be evaluated. The basic Equation (9.10) can be written in this case as follows:

$$\gamma_T T_* = 1 - \frac{F^2}{F_0^2} - \ln\left(\frac{-\ln \bar{P}}{\gamma_S S_* t + \dfrac{G^2}{G_0^2} - 1}\right). \tag{9.13}$$

For normal values of the HCF $\left(\dfrac{F^2}{F_0^2} = 1\right)$ and high values of the MWL $\left(\dfrac{G^2}{G_0^2} \ggg 1\right)$, this equation yields

$$\gamma_T T_* \approx -\ln\left(\frac{-\ln \bar{P}}{\gamma_S S_* t + \dfrac{G^2}{G_0^2}}\right). \tag{9.14}$$

The product $\gamma_T T_*$ should be always positive and, therefore, the condition

$$\gamma_S S_* t + \frac{G^2}{G_0^2} \geq -\ln \bar{P} \tag{9.15}$$

should always be fulfilled. This means that the testing time of a meaningful FOAT on a flight simulator should exceed, for the taken $\dfrac{G^2}{G_0^2}$ level, the

$$t_* = -\frac{\ln \bar{P} + \dfrac{G^2}{G_0^2}}{\gamma_S S_*} \tag{9.16}$$

threshold.

Double-Exponential Probability Distribution

Example 9.1 Let FOAT be conducted on a flight simulator or by using another suitable testing equipment for a group of individuals characterized by high HCF $\dfrac{F}{F_0}$ level at two loading conditions, $\dfrac{G_1}{G_0} = 1.5$ and $\dfrac{G_2}{G_0} = 2.5$. The tests have indicated that the critical value of the governing symptom (such as, e.g., body temperature, arterial blood pressure, oxyhemometric determination of the level of saturation of blood hemoglobin with oxygen, etc.) of the critical magnitude of, say, $S_* = 180$, has been detected during the first set of testing (under the loading condition of $\dfrac{G_1}{G_0} = 1.5$) after $t_1 = 2.0$ h of testing in 70% of individuals (so that $\bar{P}_1 = 0.3$), and during the second set of testing (under the loading condition of $\dfrac{G_2}{G_0} = 2.5$) after $t_2 = 4.0$ h of testing in 90% of individuals (so that $\bar{P}_2 = 0.1$). With these input data the formula (9.11) yields

$$\gamma_S = \frac{1}{S_*} \frac{1 - \dfrac{G_1^2}{G_0^2} - \dfrac{\ln \bar{P}_1}{\ln \bar{P}_2}\left(1 - \dfrac{G_2^2}{G_0^2}\right)}{\dfrac{\ln \bar{P}_1}{\ln \bar{P}_2}t_2 - t_2} - \frac{1}{180}\frac{1 - 2.25 - \dfrac{-1.2040}{-2.3026}(-5.25)}{\dfrac{-1.2040}{-2.3026}4 - 2}$$

$$= 0.09073 \text{ h}^{-1}.$$

Then Equation (9.4) leads to the following distribution:

$$\bar{P} = \frac{P^h(F,G,S_*)}{P_0} = \exp\left[\left(1 - \gamma_S S_* t - \frac{G^2}{G_0^2}\right)\exp\left(1 - \frac{F^2}{F_0^2}\right)\right]$$

$$= \exp\left[\left(1 - 1.8146t - \frac{G^2}{G_0^2}\right)\exp\left(1 - \frac{F^2}{F_0^2}\right)\right].$$

These results indicate particularly the importance of the HCF and that even a relatively insignificant increase in the HCF above the ordinary level can lead to an appreciable increase in the probability of human non-failure. Clearly, training and individual qualities are always important.

Let us assess now the sensitivity factor γ_T of the human error measured as his/her time to failure (to make an error). Let us check first if the condition (9.15) for the testing time is fulfilled (i.e., if the testing time is long enough to exceed the required threshold [9.16]). With $\dfrac{G_1}{G_0} = 1.5$ and $\bar{P}_1 = 0.3$, and with $\gamma_S S_* = 0.09073 \times 180 = 16.3314$, the formula (9.16) yields

$$t_* = -\frac{\ln \bar{P} + \dfrac{G^2}{G_0^2}}{\gamma_S S_*} = -\frac{\ln 0.3 + 2.25}{16.3314} = -\frac{-1.2 + 2.25}{16.3314} = -0.06405 \text{ h.}$$

The actual testing time was 2.0h (i.e., much longer). With $\dfrac{G_2}{G_0} = 2.5$ and $\bar{P}_2 = 0.1$, and with $\gamma_S S_* = 16.3314$, the formula (9.16) yields

$$t_* = -\frac{\ln \bar{P} + \dfrac{G^2}{G_0^2}}{\gamma_S S_*} = -\frac{\ln 0.1 + 6.25}{16.3314} = -\frac{-2.3026 + 6.25}{16.3314} = -0.24171 \text{ h.}$$

The actual testing time was 4.0h (i.e., much longer). Thus, the requirement (9.16) is met in both sets of tests. The formula (9.14) yields

$$\gamma_T T_* \approx -\ln\left(\frac{-\ln \bar{P}}{\gamma_S S_* t + \dfrac{G^2}{G_0^2}}\right) = -\ln\left(\frac{-\ln 0.3}{16.3314 \times 2.0 + 2.25}\right) = 3.3672$$

for the first set of testing and

$$\gamma_T T_* \approx -\ln\left(\frac{-\ln \bar{P}}{\gamma_S S_* t + \dfrac{G^2}{G_0^2}}\right) = -\ln\left(\frac{-\ln 0.1}{16.3314 \times 4.0 + 6.25}\right) = 3.4367$$

for the second set. The results are rather close, so that in an approximate analysis one could accept $\gamma_T T_* \approx 3.4$. After the sensitivity factors for the HE and SH aspects of the HF are determined, the computations for any levels of the MWL and HCF can be made.

9.9 Conclusions

The following major conclusions can be drawn from the carried-out analysis:

- The suggested DEPDF for the human nonfailure can be applied in various HITL-related aerospace problems, when human qualification and performance, as well as his/her SH are crucial, and therefore, the ability to quantify them is imperative, and since nothing

Double-Exponential Probability Distribution

and nobody is perfect, these evaluations could and should be done on the probabilistic basis.

- The MTTF is suggested as a suitable characteristic of the likelihood of a human error: if no error occurs in a long time, this time is significant; in the opposite situation it is very short.

- MWL, HCF, time, and the acceptable levels of the human health characteristic and his/her propensity to make an error are important parameters that determine the level of the probability of nonfailure of a human when conducting a flight mission or in an extraordinary situation, and it is these parameters that are considered in the suggested DEPDF.

- The MWL, the HCF levels, the acceptable cumulative human health characteristic, and the characteristic of his/her propensity to make an error should be established depending on the particular mission or a situation, and the acceptable/adequate safety level—on the basis of the FOAT data obtained using flight simulation equipment and instrumentation, as well as other suitable and trustworthy sources of information, including, perhaps, also the well-known and widely used Delphi technique (method).

- The suggested DEPDF-based model can be used in many other fields of engineering and applied science as well, including various fields of human psychology, when there is a need to quantify the role of the human factor in a HITL situation.

- The author does not claim, of course, that all the i's are dotted and all the t's are crossed by the suggested approach. Plenty of additional work should be done to "reduce to practice" the findings of this chapter, as well as those suggested in the author's previous HITL-related publications.

References

1. E. Suhir, "Human-in-the-Loop: Application of the Double Exponential Probability Distribution Function Enables One to Quantify the Role of the Human Factor," *International Journal of Human Factor Modeling and Simulation*, vol. 5, No. 4, 2017, pp. 354–377.
2. E. Suhir, *Applied Probability for Engineers and Scientists*, McGraw-Hill, New York, 1997.
3. E. Suhir, "Probabilistic Modeling of the Role of the Human Factor in the Helicopter-Landing-Ship (HLS) Situation," *International Journal of Human Factor Modeling and Simulation (IJHFMS)*, vol. 1, No. 3, 2010, pp. 313–322.

4. E. Suhir, "'Human-in-the-Loop': Likelihood of a Vehicular Mission-Success-and-Safety, and the Role of the Human Factor," Paper ID 1168. In *2011 IEEE/AIAA Aerospace Conference*, Big Sky, MT, March 5–12, 2011.
5. E. Suhir, "Likelihood of Vehicular Mission-Success-and-Safety," *Journal of Aircraft*, vol. 49, No. 1, 2012, pp. 29–36.
6. E. Suhir, "'Miracle-on-the-Hudson': Quantified Aftermath," *International Journal of Human Factors Modeling and Simulation*, vol. 4, No. 1, pp. 35–62, April 2013.
7. E. Suhir, "Human-in-the-Loop: Probabilistic Predictive Modeling, Its Role, Attributes, Challenges and Applications," *Theoretical Issues in Ergonomics Science (TIES)*, published online, vol. 16, No. 2, pp. 99–123, July 2014.
8. S. Hourlier and E. Suhir, "Designing with Consideration of the Human Factor: Changing the Paradigm for Higher Safety." In *IEEE Aerospace Conference*, Big Sky, MT, March 1–8, 2014.
9. E. Suhir, "Human-in-the-Loop (HITL): Probabilistic Predictive Modeling (PPM) of an Aerospace Mission/Situation Outcome," *Aerospace*, vol. 1, 2014, pp. 101–136.
10. E. Suhir, "Human-in-the-Loop and Aerospace Navigation Success and Safety: Application of Probabilistic Predictive Modeling." In *SAE Conference*, Seattle, WA, September 22–24, 2015.
11. E. Suhir, "Electronic Product Qual Specs Should Consider Its Most Likely Application(s)," *ChipScale Reviews*, vol. 16, No. 4, July 2012, pp. 34–37.
12. E. Suhir and L. Bechou, "Availability Index and Minimized Reliability Cost," *Circuit Assemblies*, February 2013.
13. E. Suhir, "Failure-Oriented-Accelerated-Testing (FOAT) and Its Role in Making a Viable IC Package into a Reliable Product," *Circuits Assembly*, July 2013.
14. L.P. **Goodstein**, H.B. Andersen, and S.E. Olsen (eds.), *Tasks, Errors, and Mental Models*, Taylor and Francis, Boca Raton, FL, 1988.
15. J. Reason, *Human Error*, Cambridge University Press, Cambridge, 1990.
16. E. Hollnagel, *Human Reliability Analysis: Context and Control*, Academic Press, London, 1993.
17. E. **Hollnagel**, "Context, Cognition and Control." In Y. Waern (ed.), *Co-operative Process Management, Cognition and Information Technology*, Taylor and Francis Group, New York and London, 1998, pp. 27–52.
18. S.G. Archer and L. Allender, "New Capabilities in the Army's Human Performance Modeling Tool." In *Proceedings of the Advanced Simulation Technologies Conference*, Seattle, WA, 2001.
19. R. Jagacinski and J. Flach, *Control Theory for Humans*, Lawrence Erlbaum Associates, Mahwah, NJ, 2002.
20. K.A. Gluck and R.W. Pew (eds.), *Modeling Human Behavior with Integrated Cognitive Architectures*, Lawrence Erlbaum Associates, Mahwah, NJ, 2005, pp. 307–350.
21. M.R. Lehto and J.R. Buck, *Introduction to Human Factors and Ergonomics for Engineers*, Lawrence Erlbaum Associates, Taylor and Francis Group, New York and London, 2008.
22. J. **Orasanu**, L. Martin, and J. Davison, "Errors in Aviation Decision Making: Bad Decisions or Bad Luck?" In *Proceedings of the 4th Conference on Naturalistic Decision Making*, 1998.

23. K. Leiden, J.W. Keller, and J.W. French, "Context of Human Error in Commercial Aviation," Technical Report, MicroAnalysis and Design, Boulder, CO, 2001.

24. N. Leveson, "The Role of Software in Recent Aerospace Accidents." In *Proceedings of the 19th International Systems Safety Conference, System Safety Society*, Unionville, VA, 2001.

25. A. Hobbs, "Human Factors: The Last Frontier in Aviation Safety," *International Journal of Aviation Psychology*, vol. 14, 2004, pp. 335–341.

26. D.C. Foyle and B.L. Hooey, *Human Performance Modeling in Aviation*, CRC Press, Boca Raton, FL, 2008.

27. R. **Amalberti**, *"Anticipation."* In M. de Montmollin (ed.), *Vocabulaire de l'Ergonomie*, Toulouse, Octares, 1995.

28. P. Denecker, "Les composantes symboliques et subsymboliques de l'anticipation dans la gestion des situations dynamiques," *Le Travail Humain*, vol. 62, No. 4, 1999, pp. 363–385.

29. J.M. Hoc, R. Amalberti, and G. Plee, "Vitesse du processus et temps partagé: planification et concurrence attentionnelle," *L'année Psychologique*, vol. 100, No. 4, 2000, pp. 629–660.

30. C. Mundutéguy and F. Darses, "Perception et anticipation du comportement d'autrui en situation simulée de conduite automobile." In *Le Travail Humain*, vol. 70, No. 1, 2007, p. 32, Paris: Presses Universitaires de France.

31. A. Kaddoussi, N. Zoghlami, H. Zgaya, S. Hammadi, and F. Bretaudeau, "A Preventive Anticipation Model for Crisis Management Supply Chain." In *2011 IEEE International Conference on Systems, Man, and Cybernetics*, Anchorage, AK, October 9–12, 2011, pp. 109–114.

32. S. **Lini**, C. Bey, S. Hourlier, B. Vallespir, A. Johnston, and P.-A. Favier, "Anticipation in Aeronautics: Exploring Pathways toward a Contextualized Aggregative Model Based on Existing Concepts." In D. de Waard, K. Brookhuis, F. Dehais, C. Weikert, S. Röttger, D. Manzey, S. Biede, F. Reuzeau, and P. Terrier (eds.), *Human Factors: A View from an Integrative Perspective*, 2012.

33. S. **Lini**, P.-A. Favier, J.-M. André, S. Hourlier, C. Bey, and D. Vallespir, and B. Baracat, "Influence of Anticipatory Time Depth on Cognitive Load in an Aeronautical Context," *Le Travail Humain*, vol. 78, No. 3, 2012, pp. 193–215, Presses Universitaires de France, Paris.

34. E. Suhir, C. Bey, S. Lini, J.-M. Salotti, S. Hourlier, and B. Claverie, "Anticipation in Aeronautics: Probabilistic Assessments," *Theoretical Issues in Ergonomics Science*, published online, June 2014.

35. F. Restle and J. Greeno, *Introduction to Mathematical Psychology*, Addison Wesley, Reading, MA, 1970.

36. S. Baron, D.S. Kruser, and B. Huey (eds.), *Quantitative Modeling of Human Performance in Complex Dynamic Systems*, National Academy Press, Washington, DC, 1990.

37. T.A. Polk and C.M. Seifert, *Cognitive Modeling*, MIT Press, Cambridge, MA, 2002.

38. C.D. Wickens, "Multiple Resources and Performance Prediction," *Theoretical Issues in Ergonomic Science*, vol. 3, No. 2, 2002, pp. 159–177.

39. W.K. Estes, "Traps in the Route to Models of Memory and Decision," *Psychnomic Bulletin and Review*, vol. 9, No. 1, 2002, pp. 3–25.

40. R.G. Sargent (2004) "Verification and Validation of Simulation Models." In M.E. Kuhl, N.M. Steiger, F.B. Armstrong, and J.A. Joins (eds.), *37th Winter Simulation Conference*; T.B. Sheridan and W.R. Ferrell, *Man-Machine Systems: Information, Control, and Decision Models of Human Performance*, MIT Press, Cambridge, MA, 1974.

41. A. Tversky and D. Kahneman, "Judgment under Uncertainty: Heuristics and Biases," *Science*, vol. 185, 1974, pp. 1124–1131.

42. D. Kahneman, P. Slovic, and A. Tversky (eds.), *Judgment under Uncertainty: Heuristics and Biases*, Cambridge University Press, 1982.

43. J. Rasmussen and O.M. Pedersen, "Human Factor in Probabilistic Risk Analysis and Risk Management." In *Operational Safety of Nuclear Power Plants*, vol. 1, International Atomic Energy Agency, Vienna, 1984.

44. B.F. Gore and J.D. Smith, "Risk Assessment and Human Performance Modeling: The Need for an Integrated Approach," *IJHFMS*, vol. 1, No. 1, 2006, pp. 119–139.

45. E. Suhir, "Helicopter-Landing-Ship: Undercarriage Strength and the Role of the Human Factor," *ASME OMAE Journal*, vol. 132, No. 1, 2009.

46. E. Suhir, "Remaining Useful Lifetime (RUL): Probabilistic Predictive Model," *International Journal of Prognostics and Health Management*, vol. 2, No. 2, 2011.

47. E. Suhir, "Assuring Aerospace Electronics and Photonics Reliability: What Could and Should Be Done Differently." In *2013 IEEE Aerospace Conference*, Big Sky, MT, March 2013.

48. E. Suhir and R.H. Mogford, "'Two Men in a Cockpit': Probabilistic Assessment of the Likelihood of a Casualty if One of the Two Navigators Becomes Incapacitated," *Journal of Aircraft*, vol. 48, No. 4, 2011, pp. 1309–1314.

49. J.-M. Salotti and B. Claverie, "Human System Interactions in the Design of an Interplanetary Mission." In D. De Waard, K. Brookhuis, F. Dehais, C. Weikert, S. Röttger, D. Manzey, S. Biede, F. Reuzeau, and P. Terrier (eds.), *Human Factors: A View from An Integrative Perspective*, 2012, pp. 205–212. *Proceedings HFES Europe Chapter Conference Toulouse 2012*, Downloaded from http://hfes-europe. org.

50. J.-M. Salotti, "Revised Scenario for Human Missions to Mars," *Acta Astronautica*, vol. 81, 2012, pp. 273–287.

51. J.-M. Salotti and E. Suhir, "Manned Missions to Mars: Minimizing Risks of Failure," *Acta Astronautica*, vol. 93, 2014, pp. 148–161.

52. J.-M. Salotti and E. Suhir, "Some Major Guiding Principles for Making Future Manned Missions to Mars Safe and Reliable." In *IEEE Aerospace Conference*, Big Sky, MT, 2014.

53. J.-M. Salotti, R. Hedmann, and E. Suhir, "Crew Size Impact on the Design, Risks and Cost of a Human Mission to Mars." In *IEEE Aerospace Conference*, Big Sky, MT, 2014.

54. E. Suhir, "Three-Step Concept in Modeling Reliability: Boltzmann-Arrhenius-Zhurkov (BAZ) Physics-of-Failure-Based Equation Sandwiched between Two Statistical Models," *Microelectronics Reliability*, October 2014.

55. E. Suhir, "Aerospace Electronics-and-Photonics Reliability Has to Be Quantified to Be Assured." In *AIAA SciTech Conference*, San Diego, CA, January 2016.

56. E. Suhir, L. Bechou, and A. Bensoussan, "Technical Diagnostics in Electronics: Application of Bayes Formula and Boltzmann-Arrhenius-Zhurkov Model," *Circuit Assembly*, December 3, 2012.

57. E. Suhir and A. Bensoussan, "Quantified Reliability of Aerospace Optoelectronics." In *SAE Aerospace Systems and Technology Conference*, Cincinnati, OH, September 23–25, 2014.

58. E. Suhir, A. Bensoussan, G. Khatibi, and J. Nicolics, "Probabilistic Design for Reliability in Electronics and Photonics: Role, Significance, Attributes, Challenges." In *International Reliability Physics Symposium*, Monterey, CA, 2014.

59. E.J. Gumbel, *Statistics of Extremes*, Columbia University Press, New York (2nd edition), 1958.

60. E.J. **Gumbel**, "Multivariate Extremal Distributions," *Bulletin de l'Institut International de Statistique*, vol. 37, 1960, pp. 471–475.

61. C. Bundesen, "A Theory of Visual Attention," *Psychological Review*, vol. 97, 1990, pp. 523–547.

62. D. Hamilton and C. Bierbaum, "Task Analysis/Workload (TAWL)—A Methodology for Predicting Operator Workload." In *Proceedings of the Human Factors and Ergonomics Society 34th Annual Meeting*, Santa Monica, CA, 1990, pp. 1117–1121.

63. A.D. Baddeley, "Working Memory," *Science*, vol. 255, 1992, pp. 556–559.

64. P.A. Hancock and J.K. Caird, "Experimental Evaluation of a Model of Mental Workload," *Human Factors: Journal of the Human Factors and Ergonomics Society*, vol. 35, No. 3, 1993, pp. 413–429.

65. M.R. Endsley, "Toward a Theory of Situation Awareness in Dynamic Systems," *Human Factors*, vol. 37, No. 1, 1995, pp. 85–104.

66. K.A. Ericsson and W. Kintsch, "Long Term Working Memory," *Psychological Review*, vol. 102, 1995, pp. 211–245.

67. J.M. **Cellier**, "Exigences et gestion temporelle dans les environnements dynamiques." In J.M. Cellier, V. De Keyser, and C. Valot (eds.), *La gestion du temps dans les environnements dynamiques*, Presses Universitaires de France, Paris, 1996, pp. 20–48.

68. M.R. Endsley and D.J. Garland (eds.), *Situation Awareness Analysis and Measurement*, Lawrence Erlbaum Associates, Mahwah, NJ, 2000, pp. 3–32.

69. C. Lebiere, "A Theory Based Model of Cognitive Workload and Its Applications." In *Proceedings of the 2001 Interservice/Industry Training, Simulation and Education Conference*, Arlington, VA, NDIA, 2001.

70. A. Kirlik, "Human **Factors** Distributes Its Workload." In Review of E. Salas (ed.), *Advances in Human Performance and Cognitive Engineering Research. Contemporary Psychology*, vol. 48, No. 6, 2003, pp. 766–769.

71. D.E. Diller, K.A. Gluck, Y.J. Tenney, and R. Godfrey, "Comparison, Convergence, and Divergence in Models of Multitasking and Category Learning, and in Architectures Used to Create Them." In K.A. Gluck and R.W. Pew (eds.), *Modeling Human Behavior with Integrated Cognitive Architectures*, Lawrence Erlbaum Associates, Mahwah, NJ, 2005, pp. 307–350.

72. N.A. Taatgen, "A Model of Individual Differences in Learning Air Traffic Control." In *Proceedings of the 4th International Conference on Cognitive Modeling*, Lawrence Erlbaum Associates, Mahwah, NJ, 2001, pp. 211–216.

73. S.K. Card, T.P. Moran, and A. Newell, *The Psychology of Human-Computer Interaction*, Lawrence Erlbaum Associates, Hillside, NJ, 1983.

74. W. Friedman, *About Time: Inventing the Fourth Dimension*, MIT Press, Cambridge, MA, August 9, 1990.

Index

A

Accelerated tests (ATs), 74–77
  categories, 76
Accident-free flight, probability of, 168
Activation energy, 93
Aeroflot Tupolev Tu-124, 156
Aerospace engineering, probabilistic
    modeling approach in, 1–7
Aerospace mission outcome,
    probabilistic assessment of,
    109–127
Airbus A320, 143
Air France Flight 152, 157
Algebra of events, 10–22
ALM Flight 980, 156
Angara Airlines Flight 5007, 155
Anticipation
  defined, 179
  probabilistic assessment of STA
    success from predetermined
    LTA
    DEPDF, 187–190
    LTA- and STA-related HCF, 191
    LTA- and STA-related MWL,
    190–191
  probabilistic modeling, in aviation,
    177 192
  probability that anticipation time
    exceeds certain level, 181–185
  probability that duration of
    anticipation process exceeds
    available time, 185–187
  role and attributes, 178–180
Anticipation Support for Aeronautical
    Planning (ASAP) project, 180
*Applied Probability for Engineers and
    Scientists* (book), 7
Applied probability fundamentals
  algebra of events, 10–22
  continuous random variables
    Bayes formula, 31–38
    beta-distribution, 45–51

    exponential distribution, 39–40
    normal (Gaussian) distribution,
    40–42
    probability characteristics, 29–31
    Rayleigh distribution, 42–44
    uniform distribution, 38–39
    Weibull distribution, 44–45
  discrete random variables
    Poisson distribution, 26–29
    probability characteristics, 22–26
  extreme value distributions, 54–56
  functions of random variables,
    51–54
  random events, 9–10
Arrhenius' model, 81, 93
Associativity property, 11
Auxiliary power unit, 143

B

Bayes formula, 17–18
  application, 99
  for continuous random variables,
    31–32
  as technical diagnostics tool, 33–38,
    92–93
Bernoulli trials, 25
Beta-distribution, 45–46
  application, 103
  as suitable reliability update tool
    (step 3), 98–99
  for updating reliability information,
    46–51
Binomial distribution, 15–17, 25
Bloomberg, Michael, 145
Boeing 377, 157
Boeing 737, 155
Boeing 767-260ER, 156
Boltzmann–Arrhenius' equation, 4
Boltzmann–Arrhenius–Zhurkov (BAZ)
    model, 3, 4, 46, 78, 116
  application, 99–102

219

220   *Index*

Boltzmann–Arrhenius–Zhurkov (BAZ)
   model (*cont.*)
   multiparametric BAZ model, 85–87
   physics-of-failure-based BAZ model,
     93–98
   possible way to quantify reliability,
     84–85
Boltzmann's constant, 81
Boltzmann statistics, 93
Burn-in testing (BIT), 75, 76
Bush, George W., 145

C

Casualty, 161, 166–167
   effect of time of operation on
     likelihood of, 172–174
   probability of, 168
Centered random variable, 23
Central limit theorem, 41
Certain event, 9
Choosy bride problem, 19
Clausius, 39
Coefficient of variation
   of continuous random variable, 30
   of exponential distribution, 39
   of Rayleigh distribution, 42
   of uniform distribution, 39
Cognitive overload, 162, 190
Commutativity property, 11
Complementary event, 11
Complete space, 10
Composition/convolution of
    distributions, 52, 60
Conditional probability, 14
Condition of normalization, 23
Conjugate distributions, 46
Continuous random variables
   Bayes formula for, 31–32
     as technical diagnostics tool,
      33–38
   beta-distribution, 45–46
     for updating reliability
      information, 46–51
   exponential distribution, 39–40
   normal (Gaussian) distribution,
     40–42
   probability characteristics, 29–31
   Rayleigh distribution, 42–44

   uniform distribution, 38–39
   Weibull distribution, 44–45
Control level models, 180
Cost of reliability, 87–89
Crack growth equations, 4
Cumulative distribution function
   of normal distribution, 40
   of uniform distribution, 38
   of Weibull distribution, 45
Cumulative probability distribution
    function, 29
   for extreme vertical ship velocity,
     64–65
   for relative vertical velocity, 66

D

de Havilland Australia DHA-3 Drover
    VH-DHA, 157
Delphi effort, 6
Design for reliability (DfR), of
    electronics systems, 80–81
Discovery test, *see* Highly accelerated
    life testing (HALT)
Discrete random variables, 23
   Poisson distribution, 26–29
   probability characteristics, 22–26
Distribution function, 22
Distributivity property, 11
Divi Divi Air Flight 014, 155
Double-exponential probability
    distribution function (DEPDF)
   for evaluation of likelihood
     of human nonfailure in
     emergency situation, 134–137
   for human nonfailure, 111–114,
     195–213
     consideration of time/human
      error/state of health, 202–204
     human capacity factor (HCF),
      205–206
     introduced DEPDF makes
      physical sense, 206–207
     mental workload (MWL),
      204–205
     possible FOAT-based procedure,
      209–212
     underlying reliability physics,
      207–209

Index

likelihood of vehicle operator failure to performing duties, 109, 164–166
probability of short-term anticipation (STA) success, 187–190

E

Elementary events, 10
ELXSI 6400 computer, 59
Empty space, 10
End point accelerated testing, 76
Entropy, 39
Equipment (instrumentation) failure rate, 116–117
Erlang's generalized distribution of order 2, 54
Error function, 40–41
Ethiopian Airlines Flight 961, 156
Event, 9
Experiment (trial), 9
Exponential distribution, 39–40
Extreme value distributions (EVD), 4, 54–56
Eyring's equation, 4

F

Failure-oriented-accelerated testing (FOAT), 6, 76, 77, 78, 111, 133, 200
to establishing DEPDF, 209–212
and HALT, 79–80
and qualification tests (QTs), 89
Fatal casualty, 161
Figures of merit (FOM), 126
Flow of events, 26
Flying Tiger Line Flight 923, 156
Frequency/statistical probability, of the event, 9

G

Garuda Indonesia Flight 421, 155
Gaussian distribution, 24, 40–42
Goodness-of-fit criteria, 24

H

Harten, Patrick, 144
Helicopter-landing-ship (HLS) and human factor role, 57–70

allowable landing velocity
when landing on ship deck, 66–68
when landing on solid ground, 65–66
probability distribution function for extreme vertical velocity of ship's deck, 64–65
probability of safe landing on ship's deck, 68–69
probability that duration of landing exceeds duration of lull, 63–64
probability that operation time exceeds certain time level, 59–63
Higgins, Kitty, 144
Highly accelerated life testing (HALT), 73, 75, 78
and failure-oriented-accelerated testing (FOAT), 79–80
Human capacity factor (HCF), 5, 110–111, 126, 205–206
of flight attendant, 147–149
hypothetical for flight 111 pilot, 154
LTA- and STA-related HCF, 191
and mental workload (MWL), 133–134
needed to satisfactorily cope with high MWL, 137
probability of casualty as function of, 161
required extraordinary (off-normal) vs. ordinary (normal) HCF level, 170–172
Human Engineering for Aerospace Laboratory (HEAL) effort, 180
Human-equipment-environment interaction, 121–125
Human factor role, helicopter-landing-ship (HLS) situation and, 57–70
Human-in-the-loop (HITL) events, 6
Human nonfailure
computed probabilities of, 148, 155
DEPDF of, 111–114
Human performance failure rate, 117–119

I

Impossible event, 9
Independent events, 15

Index

Instrumentation, failure rate of, 116–117
Intensity of the flow, 26, 27
Inverse power law, 4
Inverse probabilities, 18, 34

K

Kurtosis, 24

L

Landing Period Designator, 59
Landing velocity, 65–68
Laplace function, 40–41, 61
Law of probability distribution, the, 22
Leakage current, 95–96
Leibnitz, Gottfried, 132
Leptokurtic distribution, 24
Likelihood of vehicular mission success and safety, 114–115
Loew, Stephan, 150
Long-term anticipation (LTA), 179–180; *see also* Anticipation
Lull time, 63–64

M

Mathematical expectation, *see* Mean (value)
McDonnell Douglas DC-9-33CF, 156
McDonnell Douglas MD-11 airliner, 149–150
Mean (value)
  of beta-distribution, 45
  of continuous random variable, 30
  of discrete random variable, 23
  of exponential distribution, 39
  of Rayleigh distribution, 42
  of uniform distribution, 39
  of Weibull distribution, 45
Mean square deviation, *see* Standard deviation
Mean time to failure (MTTF), 111, 116–117
Mental workload (MWL), 5, 110–111, 125–126, 133–134, 137, 161–163, 178, 190–191, 204–205
Miner–Palmgren's rule, 4

Minimum total cost, of achieving reliability, 88
"Miracle-on-the-Hudson" event, 5, 131–158
  double-exponential probability distribution function (DEPDF), 134–137
  flight attendant's hypothetical HCF estimate, 147–149
  flight segments (events), 145
  HCF needed to satisfactorily cope with high MWL, 137
  incident, 142–145
  and MWL and HCF roles, 133–134, 137
  operation time vs. "available" landing time, 138–142
  other reported water landings of passenger airplanes, 155–157
  quantitative aftermath, 145–146
  Sullenberger, Captain, 143–145, 147
    computed probabilities of human nonfailure, 148
    hypothetical HCF, 148
  UN-shuttle flight
    crash, 149–153
    pilot's hypothetical HCF, 154–155
    segments (events) and crew errors, 153–154
Multiparametric BAZ model, 85–87
Multiplication rule of the probabilities, 14
Mutually exclusive events, 11–12, 15

N

"Newborn babies," 26
Nondiscrete random variables, *see* Continuous random variables
Normal (Gaussian) distribution, 24, 40–42
Northwest Orient Airlines Flight 2, 157

O

Obama, Barack, 145
Objective likelihood, 9, 23
One-dimensional Poisson's distribution, 26

Index

223

Operation time vs. "available" landing time, 138–142

Opposite/complementary event, 11

Ordered sample, 23

Outcomes, 9

P

Pan Am Flight 6, 157

Pan Am Flight 526A, 157

Pan Am Flight 845/26, 157

Paterson, David, 145

Peck's equation, 4

Physics-of-failure-based BAZ model, and three-step concept (TSC), 93–98, 99–102

Pilot-in-charge (PIC), 161

Platykurtic distribution, 24

Poisson distribution, 26–29

Poisson flow/Poisson traffic, 27

Posterior/inverse probabilities, 18, 34

Predictive modeling, 77, 78

Principle of practical confidence, 78

Prior probabilities, 18

Probabilistic aerospace electronics reliability engineering, fundamentals of

accelerated testing, 74–77

Boltzmann–Arrhenius–Zhurkov (BAZ) model

multiparametric BAZ model, 85–87

possible way to quantify reliability, 84–85

three-step concept (TSC), 93–98, 99–102

design for reliability (DfR) of electronics systems, 80–81

FOAT as extension of HALT, 79–80

practices, 73–74

probabilistic design for reliability (PDfR) concept

and principles (10 commandments), 77–78

two simple models, 81–84

qualification tests (QTs), 89

three-step concept (TSC), 89–103

total cost of reliability, minimization, 87–89

Probabilistic assessment of aerospace mission outcome, 109–127

DEPDF of human nonfailure, 111–114

equipment failure rate, 116–117

human capacity factor (HCF), 110–111, 126

human performance failure rate, 117–119

imperfect human vs. instrumentation, 121–125

mental workload (MWL), 110–111, 125–126

vehicular mission success and safety, 114–115

Weibull law, 119–120

Probabilistic design for reliability (PDfR) concept, 6, 197

and principles (10 commandments), 77–78

two simple models, 81–84

Probabilistic predictive modeling (PPM), 6

Probabilistic risk analysis (PRA) techniques, 110, 131

Probabilistic risk management (PRM), 140

Probability characteristics

continuous random variables, 29–31

discrete random variables, 22–26

Probability density function, 29

of beta-distribution, 46

of extreme random value, 56

of normal distribution, 40

of random variable, 51

Probability distribution function

for extreme vertical velocity of ship's deck, 64–65

landing velocity, 65–66

Probability distribution law, 22

Probability integral, 40–41

Probability of the event, 9

Product development testing (PDT), 75, 76

Prognostics and health management (PHM), 6

Q

Qualification tests (QT), 75, 76

next-generation, 89

224 *Index*

R

Ram air turbine (RAT), 143
Random events, 9–10
Random motion, 26
Random phenomena, 9
Random variables, 22–23
    functions of, 51–54
Rayleigh distributions, 42–44, 60,
    138–139, 181
Reliability improvement cost, 87–89

S

Safety factor (SF), 30, 60, 80, 134–135
Sample variance, 24
Short-term anticipation (STA), 179; *see
    also* Anticipation
Situation awareness, 133, 162–163
Skewed left, 24
Skewed right, 24
Skewness, 24
Skiles, Jeffrey B., 143–144
Solenta Aviation Antonov An-26
    freighter, 155
Square root of variance, *see* Standard
    deviation
Standard deviation (standard)
    of continuous random variable, 30
    of exponential distribution, 39
    of random variable, 24
State-of-health (SH) characteristic, 5
Statistics, 10
Stress-free activation energy, 97
Stress–strength interference model, 4
Sullenberger, Chesley B. (Captain),
    143–145, 147
Swissair Convair CV-240 HB-IRW, 157
Swissair "UN-shuttle" disaster (1998),
    131, 149–155
Symptoms of faults (SoFs), 92

T

Task management, 125, 133, 163
Technical diagnostics tool
    Bayes formula as, 33–38, 92–93
Three-step concept (TSC), in modeling
    aerospace electronics reliability

background, 90–91
Bayes formula
    application, 99
    as technical diagnostics tool
        (step 1), 92–93
BAZ equation
    application, 99–102
    as physics-of-failure tool (step 2),
        93–98
beta-distribution
    application, 103
    as reliability update tool (step 3),
        98–99
incentive/motivation, 89–90
Tossing of a coin, 26
Total probability formula, 15
Trial, 9
Tuninter Flight 1153, 155
"Two Men in a Cockpit"
    accident occurs when one of the
        pilots fails
    and casualty occurs if both
        fails, to performing duties,
        166–167
    effect of time of operation on
        likelihood of casualty, 172–174
    probability of casualty if one of
        pilots becomes incapacitated,
        168–170
    required extraordinary (off-normal)
        vs. ordinary (normal) HCF
        level, 170–172

U

Uncertainties, 58, 178, 180
Uniform distribution, 38–39
Unimodal (single-peak) distribution,
    24
"Unity," 23
UN-shuttle flight crash, 131
    crash, 149–153
    flight 111 pilot's hypothetical HCF,
        154–155
    segments (events) and crew errors,
        153–154
Urn problem, 25–26
US Airways Flight 1549, 142–143
    events, 145, 146

Index

225

V

Variance
    of beta-distribution, 45
    of continuous random variable, 30
    of exponential distribution, 39
    of random variable, 23
    of Rayleigh distribution, 42
    of uniform distribution, 39
    of Weibull distribution, 45
Vehicular mission, success/failure of, 114–115
Vertical take-off and landing (VTOL) situation, 59

W

Water landings of passenger airplanes, 155–157
Weakest link model, 4
Weibull distribution, 44–45, 109, 114, 119–120, 140

Z

Zimmermann, Urs (Captain), 150

Index 225

V

Variance
   of beta-distribution, 45
   of continuous random variable, 30
   of exponential distribution, 39
   of random variable, 23
   of Rayleigh distribution, 42
   of uniform distribution, 39
   of Weibull distribution, 45
Vehicular mission, success/failure of, 114–115
Vertical take-off and landing (VTOL) situation, 59

W

Water landings of passenger airplanes, 155–157
Weakest link model, 4
Weibull distribution, 44–45, 109, 114, 119–120, 140

Z

Zimmermann, Urs (Captain), 150